AF291252

UNDERWATER ROBOTICS

Engineering and Applications

UNDERWATER ROBOTICS

Engineering and Applications

Commodore Amit Ray, PhD

First published in 2025 by
PENTAGON PRESS LLP
206, Peacock Lane, Shahpur Jat
New Delhi-110049, India
Contact: 011-26490600

Typeset in AGaramond, 11.5 Point
Printed at Aegean Offset Printers, Greater Noida

ISBN 978-81-993527-1-1

www.pentagonpress.in

Admiral AK Chatterji Fellowship

Born in Dhaka on 22 November 1914, the late Admiral Adhar Kumar Chatterji, PVSM, AVSM, IN (Retd), was the first officer to hold the rank of Admiral in the Indian Navy. He was one of the first Indian cadet-entry officers to join the Royal Indian Navy (RIN) in 1933 and was commissioned on 01 September, 1935. He fought in the Second World War, in both, the Atlantic Ocean and the Indian Ocean, participating in the Burma Campaign while in command of the fleet minesweeper HMIS *Kathiawar*. In 1947, having graduated from the Royal Navy's Staff Course in England, he was appointed the Director of Naval Plans and Intelligence at Naval Headquarters, where he was instrumental in formulating the blueprint for the navy of independent India. Although budgetary constraints precluded his plan from fructifying in its entirety, much of the credit for providing the shape and form of the modern Indian Navy rests with him. From November of 1950 to January of 1953, he served as the Naval Adviser in the High Commission of India in London, before returning to India and assuming command of the Indian Navy's pride and joy — *the light-cruiser, INS Delhi.* He was thereafter appointed the Commodore-in-Charge, Bombay, an appointment that gradually evolved into that of the contemporary 'Flag Officer Commanding-in-Chief Western Naval Command'. Shortly after the completion of the Senior Leadership Course at the Imperial Defence College, London, he was promoted to the rank of Rear Admiral and appointed as the Deputy Chief of the Naval Staff (DCNS) on 08 February 1958. *In May 1962, he assumed command of the Indian Fleet and in January 1964, he was appointed the Commandant of the National Defence College, in the rank of Vice Admiral. On 01 March 1968, he was promoted* to the rank of Admiral and was appointed the seventh Chief of the Naval Staff (CNS) of the Indian Navy. Admiral AK Chatterji *retired on 28 February 1970, after 37 years of exceptionally distinguished service to the nation.*

Even after his retirement, Admiral Chatterji continued to support the several endeavours of the Indian Navy and strongly advocated intellectual development among Indian naval officers. The admiral left for his heavenly abode on 06 August 2001, but right up to the day of his passing, he remained a fervent and eloquent advocate of India's maritime growth and the development of naval and maritime leadership of the highest possible level.

To honour the multitudinous achievements of the late Admiral, his family instituted the "Admiral A K Chatterji Fellowship" in 2011 at the National Maritime Foundation (NMF) to promote his vision of effervescent yet well-grounded research that would propel maritime thinking in India. This Fellowship is offered to serving naval officers and civilian researchers, to undertake research on maritime issues of relevance.

Five research projects, each spanning a period of about eighteen months have been completed thus far under aegis of the "Admiral AK Chatterji Fellowship". Commodore Amit Ray was awarded this fellowship in May 2024, and this work is an outcome of his relentless effort and fastidious research on the subject of underwater robotics.

This volume, an outcome of the Admiral AK Chatterji Fellowship instituted at the National Maritime Foundation, is both timely and important. Around the world, armed forces are moving steadily from manned platforms towards integrated, autonomous solutions. For the Navy, Unmanned Underwater Vehicles (UUVs) hold transformative potential. By extending human reach into hazardous or inaccessible spaces below the seas, UUVs would enable persistent surveillance, oceanographic research, hydrographic survey, logistics support, mine detection and clearance, and munitions delivery—capabilities essential to maritime security in the Twenty-First century.

It is in this context that Commodore Amit Ray's contribution acquires special significance. Drawing upon his rich professional experience, including the indigenous design of the Extra-Large AUV *Jalkapi*, he presents a lucid account of the history, evolution, design, and applications of Autonomous Underwater Vehicles (AUVs) and Remotely Operated Vehicles (ROVs). His work provides not only the technical foundations of underwater robotics but also a strategic perspective on their role in national capability development.

The Indian Navy released its Integrated Unmanned Roadmap in 2021, which set forth a vision for indigenisation and self-reliance in this specialised sector. Since the early efforts of CSIR and DRDO in 2005, national initiatives have gained momentum, particularly through mechanisms such as iDEX and SPRINT. Globally, too, the trajectory of underwater robotics has been rapid, moving from tethered systems to autonomous, long-endurance platforms. Major navies are increasingly prioritising research in manned–unmanned teaming (MUM-T) and the development of Large and Extra-Large UUVs for strategic missions.

In future, large numbers of low-cost, resilient vehicles operating in concert, shifting risk from humans to machines while vastly expanding our capacity for awareness and action beneath the seas, would transform the way in which we currently look at maritime security. Achieving this vision will require continued advances in autonomy, networking, energy systems, and design reliability. This book, by laying out engineering principles of underwater robotics alongside their operational and industrial applications, provides an invaluable guide to that future.

What distinguishes this volume is its accessibility and breadth. It offers students, engineers, entrepreneurs, policymakers, and maritime professionals a clear understanding of the systems, technologies, and opportunities that define this emerging domain. By tracing both global developments and indigenous efforts, and by recommending ways to strengthen India's developmental ecosystem in this field, the author has produced a work of enduring value.

Underwater Robotics: Engineering and Applications by Commodore Ray is an important addition to the literature on marine technology. It is a book that I wholeheartedly recommend to all who are engaged in advancing our understanding and stewardship of the undersea domain.

New Delhi   **Admiral Karambir Singh**
November 2025   PVSM, AVSM (Retd)
                                                          Chairman
                                        National Maritime Foundation

Preface

The ocean depths hold immense importance for our planet's climate, marine ecosystem and energy resources. It is difficult to convey in words the opaqueness of the underwater domain experienced by a diver in the water. Without light, without sight, in depths where the human senses cannot function, we can still reach using probes that can withstand pressure, listen, measure, record, intervene and retrieve. How to do that forms the challenging and multi-disciplinary field of underwater robotics.

What can you expect from this book?

- For a tech start-up entering this field, this book provides an overview of the products, challenges, prospects, and case studies of market leaders.

- For young engineers, the book provides an alluring glimpse of the possibilities of this multi-disciplinary field.

- For labs, institutions and developers in marine robotics, this brings together details and references to point you towards details on AUV design, development, standards, operations, trials and user perspectives.

- For policy-makers and professionals in this domain, this book offers an overview of roadmaps, worldwide developments, areas requiring focus, and who is doing what in the country in this field.

- For operators and users, there are details of products, components, missions, operational examples, handling arrangements and interfaces.

- For researchers in a specific domain (AI/ML, hydrodynamics, sonars, batteries etc.), the book would introduce other technology domains, limitations, and use cases, to offer a 'systems engineering' perspective of unmanned underwater systems.

- For curious technology enthusiasts, venture capitalists or investors, the case studies, description of UUV systems, and the state-of-the-art capabilities in this exciting field would be of interest.

The Journey to this Book

My tryst with underwater robotics began in 2004 at IIT Kharagpur, where my M.Tech guide Professor Debabrata Sen, and his colleague Professor CS Kumar, were involved in development of India's first indigenous AUVs. It has been my good fortune to witness AUV development at DRDO's NSTL since 2005, thanks to Cmde N Banerjee, Dr Manu Korulla, Cmde PR Kulkarni, Dr HN Das, Dr Bharatwaja Ayyangar, Kushal Bhattacharya and Shiju John. My studies on underwater gliders after PhD in 2010 took me to interesting places and seeded some research projects at NCW IIT Delhi, which have been taken further at IIT Madras.

The Navy took about a decade more to get seriously involved. In-house design of a large AUV was the ambitious project assigned to our Navy's Submarine Design organisation by the (then) Chief of the Naval Staff, who has been kind enough to pen the foreword for this book. Admiral KB Singh's confidence in us, and the dynamic leadership of VAdm Kiran Deshmukh and RAdm Nirmal Menon, enabled our small team to take baby steps into the unmanned domain.

We named our AUV design 'Jalkapi', and my energetic team of Commanders Shashank Shankar, Alok Sahu and Venkatesh Prabhu did the entire design (for which we also filed a patent). The journey of its creation took us to R&D institutions, DRDO labs, shipyards, start-ups and mega-firms, not only in all four metro cities and hubs like Pune, Hyderabad and Bengaluru, but also to places like Durgapur, Coimbatore, Vadodara, Salem, Gandhinagar, Kochi, Tirupati, Guwahati and Goa. We were privileged to develop warm links with energetic professionals at Technology Innovation Hubs, IITs, CSIR, MoES and DRDO labs, and of course start-ups.

Trials of AUVs are exciting, and we were fortunate to dive off Dwarka with NIO's ROV, exploring ancient underwater artefacts. We witnessed several AUV lake and harbour trials in Bengaluru, Kolkata, Chennai, Kochi, Karwar and Goa, including Navy-driven iDEX projects. I am deeply grateful to Dr Pramod Maurya and his colleagues at NIO for their guidance, and for their successful development of our test-bed AUV 'Jaya'. Engagement with NIOT's

Deep Sea Technology group (Dr S Ramesh, Dr N Vedachalam and Dr M Palaniappan), and the AUV team at CMERI Durgapur led by Dr Sambhunath Nandy, have been highly enriching.

These experiences motivated me to prepare this introductory text covering AUV/ ROV applications, development roadmaps, product examples, design process, equipment, technologies and operational examples. The National Maritime Foundation's award of the Admiral AK Chatterji Memorial fellowship in 2024 was thanks to my collaborator Commander Hemanth Kumar, PhD. The NMF facilitated expert inputs, for which I am grateful to the Director General NMF, VAdm Pradeep Chauhan, and the main reviewer, VAdm Sanjay Mahindru. Their drive, erudition and guidance shaped not just this book but also several interesting conclaves by NMF on underwater uncrewed systems.

Exceptional leaders encouraged my forays into the unmanned domain. VAdm Kiran Deshmukh, VAdm SN Ghormade, VAdm SJ Singh, VAdm Rajaram Swaminathan, VAdm Krishna Swaminathan and RAdm Monty Khanna have been visionary pillars of strength. Following the dynamic RAdm Menon, his successors at the helm of SDG, RAdm Mukul Kapur and RAdm SL Saji continued to support my passion-driven ideas on AUVs, which I chased even at the risk of distraction from design responsibilities of other platforms.

Many of our efforts and ideas were stymied by convoluted systems, and biases that we are yet to outgrow. But for these, we might have become world leaders in at least few underwater solutions already. Nevertheless, the team that started in SDG with Shashank, Alok and Venkatesh, now led by Capt Pankaj Grover, has been strengthened by Rohit Shekhar, Selva Muthukumaran, Yasser, Subhash and my friend Manoj Kaimal. As new sparks join us, hopefully our dreams of a series of AUV designs, and a collaborative 'Centre of Excellence', will grow into a steady blaze of innovation, self-reliance, and new capabilities in underwater robotics.

Commodore Amit Ray

Contents

List of Figures

List of Tables

List of Abbreviations

6-DOF	Six-degrees-of-freedom
ABS	American Bureau of Shipping
ACOMMS	Acoustic Communications
ADCP	Acoustic Doppler Current Profiler
AI	Artificial Intelligence
ASME	American Society of Mechanical Engineers
ASTM	American Society for Testing and Materials
ASV	Autonomous Surface Vehicle (or Vessel)
ASW	Anti-Submarine Warfare
AUV	Autonomous Underwater Vehicle
BLDC	Brushless Direct Current (motor)
BMS	Battery Management System
C4	Command, Control, Communication and Computers
CAN	Controller Area Network
CFD	Computational Fluid Dynamics
CFRP	Carbon Fibre Reinforced Plastic
CMOS	Complementary Metal-Oxide-Semiconductor
CONOPS	Concept of Operations
CSIR	Council for Scientific and Industrial Research
CTD	Conductivity–Temperature–Depth (sensor)
CURV	Cable-Controlled Underwater Recovery Vehicle
DARPA	Defense Advanced Research Projects Agency
DDS	Data Distribution Service
DNV-GL	Det Norske Veritas- Germanischer Lloyd

DoD	Department of Defense (US)
DRDO	Defence Research & Development Organisation (MoD, GoI)
DVL	Doppler Velocity Log
EKF	Extended Kalman Filter
ELF	Extremely Low Frequency (electromagnetic waves)
ET&E	Experimental Test & Evaluation
FOG	Fibre Optic Gyro
GNC	Guidance, Navigation and Control
GoI	Government of India
GPS	Global Positioning System
GRP	Glass Reinforced Plastic
GUI	Graphical User Interface
HALE	High Altitude Long Endurance (UAV)
HAUV	Hybrid Aerial Underwater Vehicle
HIL	Hardware-In-Loop (testing or simulation)
HMI	Human Machine Interaction
H-ROV	Hybrid ROV
I-AUV	Intervention-AUV
iDEX	Innovations for Defence Excellence
IEEE	Institute of Electrical and Electronics Engineers
IMU	Inertial Measurement Unit
INU	Inertial Navigation Unit
IoT	Internet of Things
IoUT	Internet of Underwater Things
ISO	International Organisation for Standardisation
ISR	Intelligence Surveillance and Reconnaissance
IV&V	Integration, Verification, and Validation
JAUS	Joint Architecture for Unmanned Systems
KF	Kalman Filter
LARS	Launch and Recovery System
LBL	Long Baseline (acoustic communication)

LDUUV	Large Displacement Unmanned Underwater Vehicle
LiDAR	Light Detection and Ranging
Li-ion	Lithium-Ion (cell chemistry)
LiPo	Lithium-Polymer
LOS	Line-of-sight (communication)
MALE	Medium Altitude Long Endurance (UAV)
MBES	Multi Beam Echosounder
MCM	Mine Countermeasure (operations)
MEMS	Micro Electro-Mechanical Systems
ML	Machine Learning
MoD	Ministry of Defence (GoI)
MoES	Ministry of Earth Sciences (GoI)
MOOS-IvP	Mission Oriented Operating Suite—Interval Programming
MSME	Micro, Small and Medium Enterprises
NiMH	Nickel-Metal Hydride (cell chemistry)
NIO	National Institute of Oceanography (Goa)
NIOT	National Institute of Ocean Technology (Chennai)
NMEA	National Marine Electronics Association
NOAA	National Oceanic and Atmospheric Administration
ONR	Office of Naval Research
OTEC	Ocean Thermal Energy Conversion
PEM	Proton Exchange (or Polymer Electrolyte) Membrane (Fuel Cell)
PLAN	People's Liberation Army-Navy (China)
PLUSNet	Persistent Littoral Undersea Network
RF	Radio Frequency
ROS	Robot Operating System
ROV	Remotely Operated Vehicle
SAS	Synthetic Aperture Sonar
SATCOM	Satellite Communication
SBL	Short Baseline (acoustic navigation)

SE	Systems Engineering
SIoT	Subsea Internet of Things
SLAM	Simultaneous Localisation and Mapping
SNR	Signal-to-Noise Ratio
SPAWAR	Space and Naval Warfare Systems Command (US Navy)
SSS	Side Scan Sonar
TTL&R	Torpedo Tube Launch and Recovery
UAV	Unmanned (or uncrewed) Aerial Vehicle
UHR	Underwater Humanoid Robots
UKF	Unscented Kalman Filter
UMAA	Unmanned Maritime Autonomy Architecture
UMCC	Unmanned Maritime Common Control
USBL	Ultra-Short Baseline (acoustic navigation)
USV	Unmanned Surface Vehicle
UUV	Unmanned Underwater Vehicle
UUS	Unmanned Underwater Systems
UWSN	Underwater Wireless Sensor Network
UxS	Unmanned Systems (in general)
UXV	Unmanned Surface/Underwater Vehicle
VCB	Vertical Centre of Buoyancy
VCG	Vertical Centre of Gravity
VLF	Very Low Frequency (electromagnetic waves)
WHOI	Woods Hole Oceanographic Institution (US)
WPT	Wireless Power Transfer
XLUUV	Extra-Large Unmanned Underwater Vehicle

1

Underwater Robotics Applications

Marine robotics is an interdisciplinary field involving design, development, deployment and application of automated or remotely operated robotic systems engineered for use in marine environments such as oceans, seas, and inland water bodies. This field has broadened the reach and opportunities in the ocean sciences and holds promise to revolutionise maritime security. In this opening chapter we look at the types of systems, their main components, and examine their applications in scientific, industrial and defence domains.

Introduction

Underwater robotics is the specialised field of designing, developing, and utilising robotic systems for operation in submerged environments not only in oceans or seas, but also in harbours, lakes, and rivers. These are primarily categorised into Remotely Operated Vehicles (ROVs) and Autonomous Underwater Vehicles (AUVs), but also include other types such as Underwater Gliders, Hybrid vehicles (H-ROVs), Intervention-AUVs (I-AUVs) and Underwater Humanoid Robots (UHRs). Underwater robotics also includes development and integration of all mechanical, electrical, and electronic subsystems necessary for robotic function in underwater environments, such as sensors, actuators, docking aids, charging systems, and systems for launch and recovery. It also includes undersea or seabed sensors.

Oceans form the most significant proportion of our planet, but their depths remain extremely inaccessible. Satellite coverage of the planet has transformed land-based localisation and navigation, and facilitates tracking of ships at sea. However, the ocean depths continue to be impervious and opaque due to physical limitations of the electromagnetic spectrum to penetrate sea water. Underwater robotics extends human capabilities for monitoring into inaccessible underwater spaces, enabling sustained operations and data collection that would otherwise be dangerous, costly, or impossible for human divers or conventional vessels.

The potential of underwater robotics lies in the prospect of low cost, accessible and large number of vehicles working together to provide wider, deeper and long-lasting awareness of ocean depths. Driven by miniaturised electronics, compact energy sources, powerful computing hardware, and advances in Machine Learning and Artificial Intelligence, underwater robotics has matured significantly in the past two decades. These unmanned systems can serve as an enabler and force multiplier to better understand and monitor the oceans.

Unmanned and Autonomous Solutions

Unmanned surface vessels (USVs) operate on the sea surface. Unmanned underwater vehicles (UUVs) denote both autonomous vehicles (AUVs) and remote-operated vehicles (ROVs). Together, USVs and UUVs may be referred to as UXVs (where X denotes either surface or underwater).

Development of oceanographic submersibles and military submarines in the late 19th century and first half of the 20th century necessarily involved a human crew inside the vehicle to operate it. Their design was therefore driven not only by their scientific or military purpose, but also by the need to ensure safety of their crew. The challenges that were progressively overcome to reach greater depths and to evolve to unmanned systems included:

- Watertight and pressure-resistant shell: First of iron, later of higher strength steels, and (much) later, aluminium alloy, titanium alloy and composites.

- Energy storage and prime mover: From human-power to diesel engines to batteries.

- Sensors and communications: Essentially relying on sound underwater, but using radio when on surface, or if connected to the surface using cables to a mother ship.

- Ability to work: Engineering systems to launch torpedoes, raise or lower equipment, or manipulate tools (using compressed air, hydraulics, or electrical means).

- Autonomy: Linked to development of electronics, automation, and growth in computing power.

Remotely Operated Vehicles

ROVs are tethered, human-controlled vehicles used for real-time observation, inspection, intervention, and manipulation tasks. They are further subdivided by size and function (micro, mini, inspection, light work-class, heavy work-class, and specialised trenching/burial ROVs).

ROVs originated as unmanned equipment connected to the deploying ship through cable(s) that provided power, enabled communication, and enabled lowering and raising them. They were used for taking measurements, recording oceanographic data, or for taking photographs. Robert Ballard describes how he modified and used a towed equipment to take thousands of pictures of the sea bottom, trying to locate the wreck of the *Titanic* in 1985.[1] This method of deep-sea search proved to be more efficient than deploying a manned submersible, which had constraints in endurance, range and ease of deployment/recovery.

Industrial applications for oil and natural gas sectors have driven the growth of 'work-class' ROVs. These are equipped with manipulators, probes, cameras and lights, controlled by shipboard operators through the connecting umbilical (tether) that provides power and exchanges signals and data.[2]

Figure 1-1. ROVs at NIOT Chennai (Photo: Author)

---

[1] Ballard, 2021

[2] Christ, 2011

Autonomous Underwater Vehicles

AUVs are un-tethered, self-guided vehicles that operate independently using onboard navigation and control systems. Autonomous vehicles have provided greater flexibility in scientific data collection, without the constraints of an umbilical linked to a support ship, but with the attendant disadvantage of limited energy storage and the risk of losing the vehicle. Underwater localisation, navigation and communication continue to remain major challenges for reliable and coordinated operations.

The modern torpedo with its ability to seek and trail its target at high speed is also an autonomous vehicle in several ways. However, the term AUV is used to denote a more benign range of slow-moving, limited endurance vehicles, which were initially developed for oceanographic applications in the 1980s.

The proliferation of interest in robotics, AI/ML and drones among students and researchers in the past two decades provides a highly conducive climate today for marine robotics. Developments in several enabling technologies are bringing about a 'democratisation' of underwater studies, with inexpensive and capable options becoming available for components, batteries, hardware and software.

Figure 1-2. AUV 'Jaya' developed by CSIR-NIO, Goa (Photo: Author)

Access to the seas for trials, experimentation and study is restricted and expensive as compared to land or aerial applications. Nevertheless, domains that were once restricted to navies, or required expensive voyages by hiring ships, are gradually opening up to academia and tech start-up firms brimming with confidence. Companies across the world, both large and small, offer increasingly affordable solutions as well as components for commercial and scientific underwater and autonomous applications.

Other Vehicle Types

Most of the underwater robotics industry is focused on AUVs and ROVs. There other important types of UUVs in service as well, introduced below:

- **Underwater Gliders**: Energy-efficient AUVs that use buoyancy changes to move through the water in a sawtooth pattern, enabling very long-term oceanic profiling missions.[3]

- **Underwater Humanoid Robots**: UHRs mimic human movements and enhance adaptability in complex tasks.[4]

- **Hybrid Vehicles**: These are also referred as Hybrid ROVs (H-ROVs) and include Intervention-AUVs (I-AUVs). These platforms can operate both autonomously and under remote control, combining AUV efficiency for transit/surveys with ROV-style human oversight for intervention or complex manipulation.[5]

Building Blocks for Unmanned Systems

Any unmanned (uncrewed) marine vehicle, UUV as well as USV, would have the following categories of systems:

- Vehicle systems for operations (propulsion, manoeuvring, sensors, manipulators, etc.).

- Command and control systems for decision-making, mission execution and communication.

- Support systems for launch, recovery, maintenance, recharging, etc.

Technologies and Equipment

The integral technologies and aspects of equipment required for UUVs and USVs include structural design, hydrodynamics, control systems, energy storage and distribution systems, mechanical systems, navigation, communication and overall mission control (autonomous operation). Their maturity in the Indian context is represented in Figure 1-3, where the shading is used to qualitatively denote indigenous maturity (aspects needing most attention shown in the darkest shade).

---

[3]   Ray, Singh, & Seshadri, 2011

[4]   Canjun Yang, 2024

[5]   Dunbabin, 2005

Functional Capabilities

The non-availability of continuous communication while the vessel is dived requires the vehicle mission computer to be fault-tolerant and quick to detect and respond to emergencies. Sensors and components need to have adequate reliability and redundancy. Inertial navigation becomes essential for submerged operation in the absence of GPS.

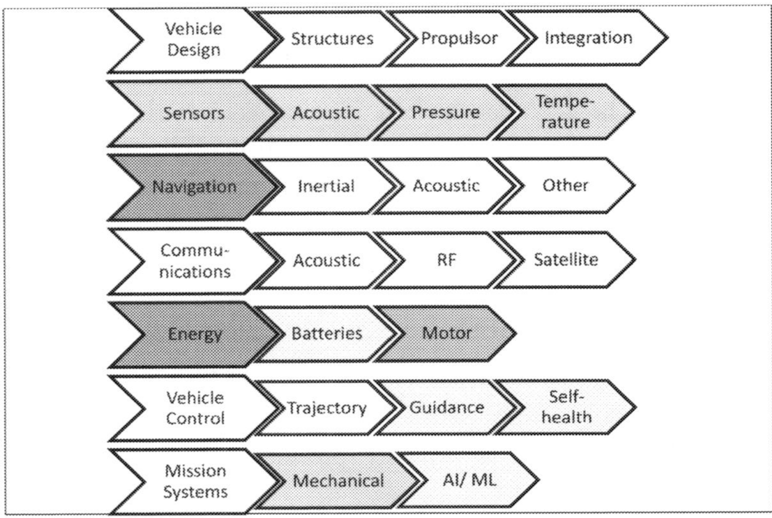

Figure 1-3. Systems and Technologies required for Unmanned Marine Vehicles
(Source: Author)

The UXV should be able to communicate at the sea surface via satellite. While underwater, suitable acoustic communication systems will be required. Acoustic sensors and communication devices require extensive testing and calibration, and are still dependent on environmental conditions. The mechanical systems required include low-power, high-torque motors, efficient thrusters/propellers and suitable manipulators for underwater use, as necessary.

Compact batteries with high-energy density are essential for adequate endurance. A suitable energy management system is required for battery monitoring and smart power distribution. Battery charging options for the vehicle would determine its endurance, turnaround time, and operational availability.

The sensors need to be compact, and should be able meet multiple requirements wherever feasible. The sensors may include lights, cameras, sonar, radar, conductivity-temperature-depth (CTD) sensors, etc., and would drive the configuration and design of the UXV.

The vehicle should be able to collect and transmit data with various autonomous protocols. It needs to be capable of scenario-based decision-making. For example, the AUV is expected to switch between different means of communication for transmitting different types of data. The UXV should be able to communicate with shore station(s), other UXVs, and specified nodes in a marine network.

Related Technologies

Related technologies that are necessary include a human interface at the command post for tracking and monitoring the UXV, with possibly a digital twin for predicting the performance of its systems/equipment. A predictive maintenance approach should be in-built to reduce equipment down time.

Several geophysical and sampling systems are used to define the seafloor topography, surface sediment, and underlying geology, which have been represented in Figure 1-4. Side-scan-sonar systems acquire information about the surface of the seafloor. Swath bathymetric systems measure the depth, or seafloor topography. Seismic sources map the underlying geologic structure, and single-beam echo-sounders map the depth at a point beneath the vessel. Sampling systems collect samples of the seafloor and can be equipped with systems to collect acoustic/optical images of the seafloor.

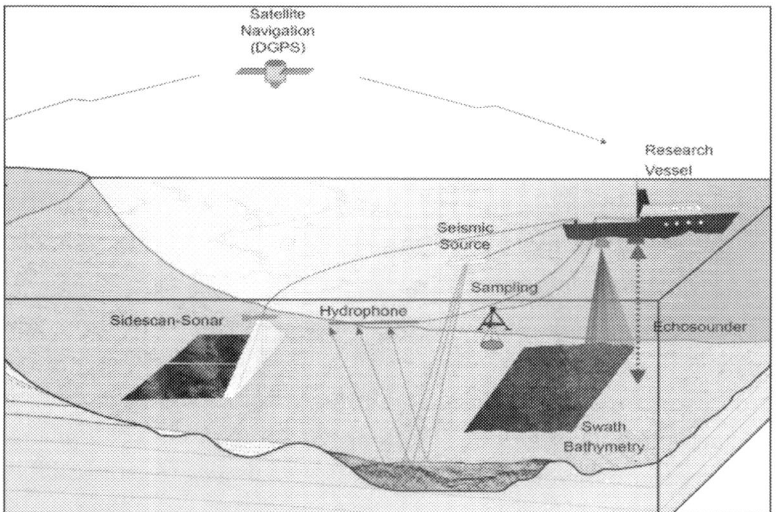

Figure 1-4. Seafloor survey and mapping systems
(Source: Woods Hole Coastal and Marine Science Centre[6])

---

[6]   https://www.usgs.gov/media/images/seafloor-mapping-systems

Scientific and Industrial Applications

Marine robotics spans UXVs as well as fixed sensors that together can be used for marine remote sensing, measurement, sampling or intervention. The enabling technologies include compact electronics, flexible stored energy, structural design to withstand hydrostatic pressure, suitable sensors, and powerful hardware and software. Growing computational power and compact sensors, actuators and energy storage solutions have fuelled a rapid growth in marine robotics worldwide over the past two decades.[7] Various applications of marine robotics are represented in Figure 1-5.

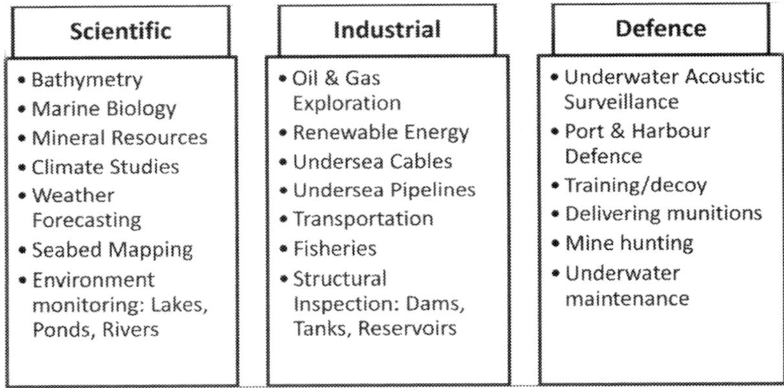

Figure 1-5. Applications of Marine Robotics (Source: Author)

Scientific Applications

Any scientific study requires data. Scientific data collection in the oceans has relied upon expensive and far too few scientific ships for exploration. Submersibles and towed sensors made forays into the depths to study the oceans, their biodiversity and resources, and to collect information about its vast and varied characteristics.

The development of uncrewed and remotely-operated solutions was driven by the need to overcome the constraints of deploying sensors and collecting data for monitoring the oceans. Unmanned underwater systems have transformed the field of marine science by enabling extended, high-resolution studies of ocean dynamics, ecosystems, and geological features. UUVs have

---

[7]   Griffiths, 2002.

proved to be invaluable for exploring areas that are too remote, deep, or hazardous for human divers.

Autonomous vehicles have provided further flexibility in scientific data collection, without the constraints of an umbilical linked to a support ship. However, these come with the attendant disadvantage of limited energy storage and the risk of losing the vehicle. Underwater localisation, navigation and communication continue to remain major challenges for reliable and coordinated operations using AUVs.

Examples of Applications

Research on hydrothermal vent ecosystems relies on AUVs capable of high-resolution mapping and in-situ chemical analysis. For example, dedicated missions have used gliders to track temperature anomalies and chemical signatures associated with vent plumes. These missions not only map vent locations but also study the unique biota that thrive in extreme conditions.

Under polar conditions, unmanned underwater systems are deployed to explore regions beneath thick ice sheets. In such environments, the ability to operate autonomously for extended periods is critical. Vehicles like the Explorer AUV have been modified to withstand low temperatures and to navigate in complex, ice-laden waters[8]—providing valuable data on climate change and polar ecosystems.

Underwater gliders, such as the Slocum Glider, are a prime example of energy-efficient scientific vehicles.[9] These systems adjust their buoyancy to glide through the water, offering endurances of months and the ability to cross entire ocean basins. Gliders have been successfully deployed to measure ocean temperatures, currents, and biogeochemical properties over extended periods.

Requirements of Scientific Missions

Scientific missions involving unmanned underwater studies typically require the features and capabilities described below:

- **High-Resolution Data Collection:** Biological imaging, water column profiles (temperature, salinity, and chemical composition), and

---

[8] Butler, 2018
[9] Ray, 2011

thorough mapping of the seafloor are required. Oceanographers use AUVs to take pictures of deep-sea creatures and gather CTD data.

- **Endurance and Range:** For many scientific missions, such as trans-oceanic surveys or hydrothermal vent monitoring, long-duration operations (from a few hours to months) are necessary to cover large areas.

- **Precision Navigation:** For mapping applications, precise localisation is essential. Multiple navigation systems must be integrated by AUVs in order to prevent drift and maintain accurate trajectories.

- **Sturdy Sensor Integration:** Research frequently necessitates the concurrent use of many sensors, such as specialized sensors for environmental factors, optical cameras for imaging, and sonar arrays for bathymetry.

- **Autonomous Adaptability:** Systems need to be able to make decisions in real time, such as modifying their sampling plan when they come across unexpected features, because the undersea world is unpredictable.

Many modern AUVs integrate multiple sensor types into a single, compact platform. For instance, vehicles developed under academic and governmental research programs can simultaneously record videos, conduct sonar mapping, and sample water chemistry. The *Explorer* AUV, developed as part of China's 863 Program, is one such example, featuring advanced navigation systems and sensor suites that can be reconfigured for different types of surveys.[10]

Recent advances have also focused on onboard data processing using machine learning algorithms. These systems can analyse sensor data in real time to detect anomalies or to optimize survey paths. For example, machine learning has been applied to process sonar data, helping to identify geological features or biological hotspots automatically.

---

[10]   Fedasiuk, 2021

Table 1-1. Challenges for UUVs in Scientific and Industrial Applications

Problem Area	Challenges	Solutions Required
Harsh Environmental Conditions	Extreme pressures, corrosive salt water, and bio-fouling can degrade system performance over time.	Materials research is essential to develop more robust, corrosion-resistant structures and coatings. Self-cleaning surfaces and bio-fouling-resistant materials could significantly extend operational lifespan.
Navigation and Positioning	Without periodic updates or fixes, errors in dead reckoning accumulate rapidly, leading to inaccurate estimation of position	Continued development of hybrid navigation systems combining DVL, inertial measurements, and acoustic positioning is necessary. Periodic surfacing to acquire GPS fixes remains a critical but energy-costly procedure. Future research may focus on improved underwater localization techniques using networked acoustic beacons or integrated optical systems.
Energy Management and Endurance	Extended missions require long battery life and efficient power management, yet battery technology remains a limiting factor.	Continued innovation in battery technology for higher energy density, and integration of energy-harvesting devices (such as harnessing ambient energy from ocean currents), can boost endurance. Research into alternative power sources, such as fuel cells or renewable energy converters, is also essential.
Communication and Data Bandwidth	Limited reliability and bandwidth of acoustic communication underwater hinders real-time data transmission.	Advances in acoustic communication protocols and the exploration of alternative modalities (e.g., optical or hybrid systems) are necessary to improve data throughput. Research into adaptive modulation schemes and error-correction techniques could further enhance underwater communications.
Data Integration and Real-Time Decision Making	Operations require real-time decision-making based on data from underwater systems.	Onboard AI algorithms that pre-process data and trigger alerts locally can help operators make timely decisions.
Robust Autonomy in Complex Environments	Navigation is required in cluttered and dynamic environments.	Enhanced sensor fusion and machine learning approaches that enable real-time obstacle avoidance and adaptive mission planning will be vital. Solutions need to integrate redundant sensors and adaptive control systems.

Industrial Applications

Industrial applications of underwater vehicles have been primarily in the oil and natural gas sector. These employ 'Work-class' ROVs, i.e., capable of undertaking physical manipulation tasks underwater. ROVs have proved to be highly effective substitutes for risky, expensive and time-taking operations by human divers (Figure 1-6).[11] Such ROVs are equipped with manipulators, probes, cameras and lights. They are remotely controlled by operators on a ship, through the connecting umbilical that provides power and exchanges signals and data.

Unmanned underwater solutions are also being used increasingly in marine construction and subsea infrastructure inspection. Forecasts project the global market for unmanned underwater vehicles to grow from $ 3.2 billion in 2023 to $ 9.2 billion by 2030, highlighting the rapidly increasing industrial demand for these systems.[12]

Requirements for Industrial Missions

Industrial missions demand systems that not only operate reliably under harsh conditions but also provide precise and actionable data for maintenance, safety and resource management.

Figure 1-6. Underwater welding
(Photo by Kjartan Lerøen using Blueye ROV[13])

---

[11]   Christ, 2011.

[12]   UUV Industry Research Report, 2024.

[13]   https://www.blueyerobotics.com ; Pantheeradiyil, 2024.

UUVs for industrial use need to have salient features for the respective roles indicated as follows:

- **Precision Inspection and Surveying**: Inspection of pipelines, subsea installations, and structural integrity of offshore platforms requires high-resolution imaging and precise navigation.

- **Robust Data Collection**: Industries need real-time, accurate data to monitor structural health, corrosion, or bio-fouling on submerged assets.

- **Long Endurance and Reliability**: Industrial operations may require continuous monitoring over long periods, sometimes spanning months. UUVs must be robust and require minimal human intervention.

- **Rapid Deployment and Versatility**: In emergency scenarios such as oil spill response or structural failure rapid deployment is critical. The systems must be versatile to adapt to a variety of tasks with minimal reconfiguration.

- **Integration with Control Systems**: Data collected by UUVs must be easily integrated into existing asset management and maintenance systems, often through wireless or wired communication interfaces.

Offshore Construction and Maintenance

Maintenance of offshore wind farms, oil platforms, and underwater cables are complex industrial tasks. ROVs are often used to perform routine inspections, corrosion monitoring, and minor repairs. These vehicles are designed with high manoeuvrability and stability in rough underwater conditions, ensuring safe operation around critical infrastructure.

ROVs and AUVs equipped with high-definition cameras and corrosion sensors are deployed to inspect turbine foundations, subsea cables, and support structures. The data collected informs maintenance schedules and enables predictive maintenance, reducing unexpected downtime and repair costs. By reducing downtime and enhancing the accuracy of inspections, unmanned underwater systems are transforming the maintenance routines of offshore installations.[14] During mandatory repairs to be carried out by human divers in low visibility conditions, underwater, ROVs and AUVs can be effective for initial survey, job monitoring and diver support.

---

[14] Xiang Liu, 2025

Pipeline and Infrastructure Inspection

Inspection of underwater pipelines is critical for safe and efficient operation of offshore oil and gas facilities (Figure 1-7). Detection of early signs of pipeline corrosion enables timely intervention to prevent potential leaks.

Commercial AUVs and ROVs are now routinely deployed for pipeline inspections, which are a major sub-set of maintenance tasks on rigs and other subsea structures.[15] Modular AUVs are equipped with high-definition cameras, side-scan sonars, and laser scanning systems to assess pipeline conditions and detect potential hazards.[16]

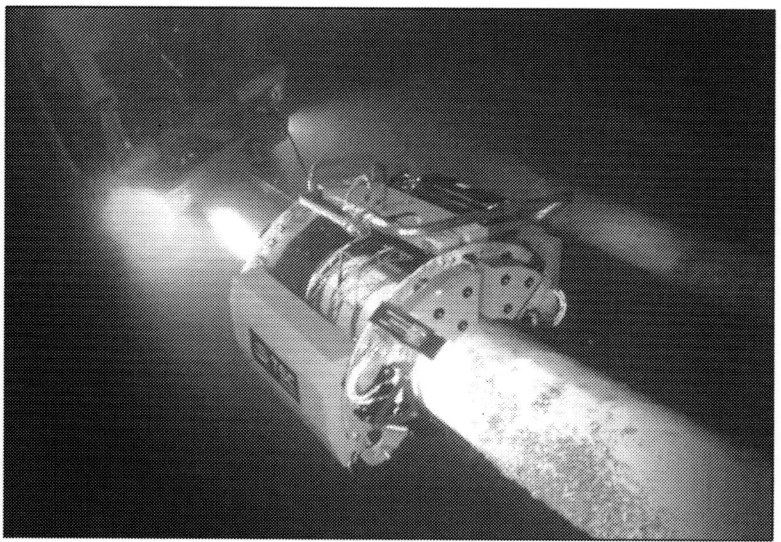

Figure 1-7. Subsea pipeline inspection by ROV (on left)
(Source: TSC Subsea[17])

Deep-Sea Mining and Resource Mapping

Industrial exploration of mineral resources on the seafloor requires precise mapping and sampling of underwater geology. Unmanned systems like AUVs equipped with advanced sonar mapping and sediment sampling tools are used to create three-dimensional maps of the seafloor and assess resource potential. Market research indicates that the UUV market in deep-sea

---

[15] Rumson, 2021

[16] Zhang, 2020

[17] https://www.tscsubsea.com/pipeline-inspections

exploration is expected to grow significantly in the coming years, driven by technological advances and lower operational costs.[18] These missions not only identify resource-rich areas but also help assess environmental impacts, ensuring sustainable mining practices.

Autonomous Intervention and Repair

Recent industrial advances include the exploration of robotic intervention systems that can perform autonomous repairs on underwater structures. While still largely in the research phase, prototypes featuring robotic manipulators and dexterous arms have been tested for tasks such as cleaning, welding, and sealing damaged areas on subsea pipelines.

Underwater Survey

Underwater survey is a basic function that forms a common and substantial part of the operational profile of most AUVs. Data collected by AUVs through underwater surveys are invaluable for a wide range of applications. This section describes how survey data are used across various domains and the purposes they serve.

Applications of Underwater Surveys

Oceanography and Climate Studies

- **Bathymetric Mapping:** Detailed maps of the seafloor help scientists understand ocean circulation patterns, sediment transport, and tectonic activity.

- **Environmental Monitoring:** Data on water temperature, salinity, dissolved oxygen, and chlorophyll concentration provide insights into climate change, marine ecosystems, and biogeochemical cycles.

- **Sub-Bottom Profiling:** Investigating subsurface structures aids in understanding sedimentary processes, earthquake hazards, and the history of ocean basins.

- **Marine Biology and Ecology:** Tracking of marine mammals (such as whales), monitoring of coral reefs, measurement of chlorophyll levels, etc.

---

[18]   UUV Industry Research Report, 2024

- **Habitat Mapping:** High-resolution imagery and sonar data enable the mapping of coral reefs, sea-grass beds, and other critical habitats.
- **Biodiversity Studies:** Surveys can help document species distribution and abundance, supporting conservation and management efforts.
- **Impact Assessments:** Monitoring environmental changes following events such as oil spills or coral bleaching provides critical data for environmental impact assessments.

Offshore Oil and Gas Exploration

- **Seafloor Characterization:** Detailed bathymetric and sub-bottom data help identify potential drilling sites, assess reservoir properties, and plan pipeline routes.
- **Risk Assessment:** Surveys can identify underwater hazards such as shallow areas, rocky outcrops, or areas of high sedimentation that may pose risks to infrastructure.
- **Regulatory Compliance:** Accurate survey data are often required to meet regulatory standards and ensure safe installation of offshore infrastructure.

Marine Infrastructure Inspection

- **Pipeline and Cable Inspection:** AUVs are used to inspect underwater pipelines, cables, and other infrastructure, detecting corrosion, damage, or bio-fouling.
- **Port and Harbour Surveys:** Detailed surveys of harbour areas support dredging operations, maintenance planning, and navigational safety.
- **Renewable Energy:** In the context of offshore wind farms and tidal energy installations, AUV surveys help assess seabed conditions and monitor structural integrity.

Societal and Environmental Benefits

- **Disaster Response:** In the aftermath of natural disasters, underwater surveys can assess damage to coastal structures and provide data for reconstruction planning.
- **Archaeological Study:** UUVs have led to significant discoveries of shipwrecks and submerged cultural heritage sites. Use of ROVs and

AUVs are now routinely used for studying underwater archaeological sites, enhancing our understanding of history (Figure 1-8).

- **Resource Management:** Data from underwater surveys support fisheries management, marine spatial planning, and environmental conservation initiatives.

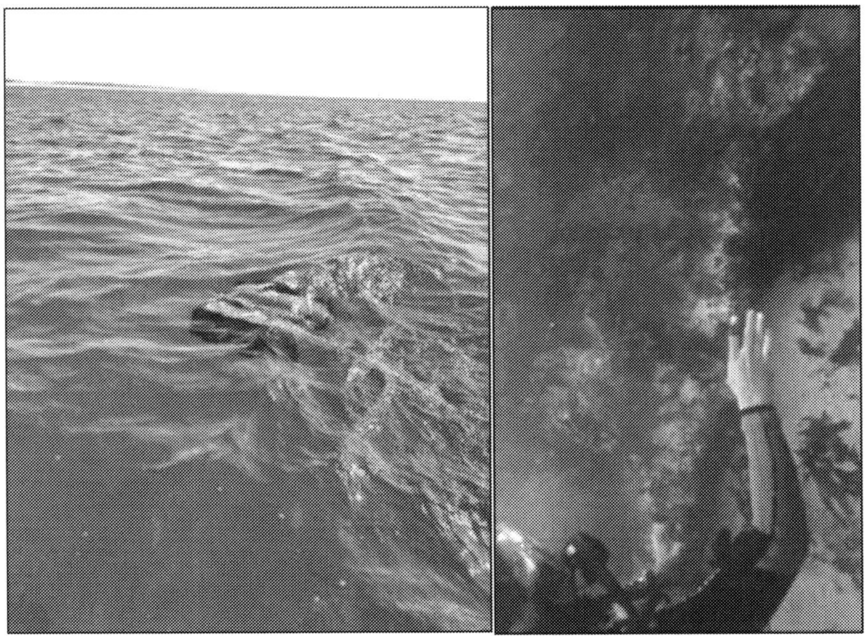

Figure 1-8. ROV by NIO Goa (left) in use off Dwarka to survey underwater stone anchors (right) on the seabed
(Source: NIO Goa)

Challenges in Underwater Survey by AUVs

While AUVs have revolutionized underwater surveys, several challenges persist that are specific to their non-tethered character. These challenges can be broadly classified into environmental, technical, and operational categories.

Environmental Challenges

- **Optical Attenuation:** Water attenuates light, reducing visibility and affecting the performance of optical cameras. Turbid waters or those with high plankton concentrations further reduce image quality.

- **Acoustic Variability:** Sound speed in water varies with temperature, salinity, and depth. These variations affect the performance of sonar systems and complicate data interpretation.

- **Pressure and Temperature Extremes:** At great depths, high pressure and low temperatures can affect both sensor performance and the structural integrity of the AUV. Equipment must be designed to withstand these harsh conditions, which increases cost and complexity.

- **Currents and Turbulence:** Ocean currents and turbulent flows can displace AUVs, affecting navigation accuracy and survey repeatability. Dynamic environments require advanced control algorithms and robust INS (Inertial Navigation Systems) to maintain course.

- **Marine Life and Debris:** Biological entities and floating debris can interfere with sensors, cause collisions, or generate false returns in sonar data.

Technical Challenges

Navigation and Positioning. Accurate navigation is paramount for geo-referencing survey data. The challenges are listed as follows:

- **GPS Limitations:** GPS signals do not penetrate water, so AUVs must rely on INS, acoustic positioning systems (such as USBL or LBL), and dead reckoning, which can accumulate errors over time.

- **Sensor Drift:** Inertial sensors can drift over long missions, necessitating periodic calibration or correction from external references.

- **Terrain Referencing:** While terrain-aided navigation can improve accuracy, it requires pre-existing high-quality bathymetric maps, which may not always be available.

Energy Management

- **Limited Power:** AUVs are typically battery-powered, and energy must be conserved for propulsion, sensor operation, data processing, and communication. Energy limitations restrict mission duration and the amount of data that can be transmitted.

- **Balancing Energy and Data Needs:** Energy conservation strategies must be balanced with the need for timely data reporting, posing a significant design challenge.

Communication Constraints

- **Bandwidth Limitations:** Underwater acoustic communication offers limited bandwidth and high latency, making real-time data transmission challenging.

- **Signal Attenuation and Multipath:** Acoustic signals are susceptible to multipath effects and attenuation, which can lead to data loss or corruption.

- **Intermittent Connectivity:** Communication links may be intermittent, necessitating robust on-board data storage and delayed data transmission strategies.

Operational and Logistical Challenges

Mission Planning and Execution

- **Complex Mission Planning:** Designing survey missions that maximize area coverage while ensuring high-resolution data collection requires detailed planning and sophisticated software tools.

- **Risk Management:** Operating AUVs in challenging underwater environments involves risks such as collisions, entanglement, or sensor malfunctions. Effective risk management protocols and redundancy in sensor systems are essential.

- **Recovery and Maintenance:** Post-mission recovery of AUVs can be challenging, especially in rough seas or remote areas. Additionally, the maintenance and repair of sophisticated AUV systems are costly and require specialized expertise.

Data Overload and Analysis

- **Volume of Data:** Modern AUVs generate massive amounts of data, which can overwhelm storage systems and make real-time analysis difficult.

- **Data Integration:** Integrating heterogeneous data streams from multiple sensors into a coherent dataset for analysis requires advanced data fusion and processing techniques.

- **Quality Control:** Ensuring data quality—such as correcting for sensor errors, filtering noise, and calibrating measurements—is an ongoing challenge.

Defence Applications

Unmanned underwater systems are increasingly critical to modern naval operations. They offer a means to conduct covert reconnaissance, mine countermeasures, anti-submarine warfare, and even offensive strikes without risking human lives. The defence sector's emphasis on stealth, endurance, and rapid responsiveness drives much of the technological innovation in Unmanned Underwater Systems.

Naval Roles and Missions

Oceanography

Navies have funded and driven research in ocean sciences, seabed mapping and physical oceanography, particularly during and after the Second World War.[19] Mapping the seabed is of obvious navigational significance, which has been steadily improving in coverage and refinement over the years, but is far from complete. Bathymetric conditions (temperature, density, salinity) of the ocean waters vary with location, season, depth and time of the day. These impact the behaviour of sound in water, thus affecting detectability and effectiveness of submarines. Understanding these phenomena in our areas of operations is crucial and requires sustained efforts. Marine robotics now offers the capability to undertake physical oceanographic measurements over extended durations, in a variety of circumstances, and enable real-time updates.

Inspection and Survey

The Navy's ships and submarines require regular maintenance cycles, involving manpower-intensive repairs and infrastructure for docking. Automated or robotic hull cleaning and inspection using underwater crawlers are becoming increasingly effective. These need to be robust and reliable in mapping and identification of defects. Developments in underwater welding and robotic non-destructive testing would enable quick and cost-effective underwater repairs, possibly avoiding emergency docking of ships.

Surveillance and Reconnaissance

Area monitoring and surveillance are the most pervasive of maritime security requirements. The length of the 11,099-km long coastline of India is about

---

[19] Weir, 2001

three-fourths of its land borders. Continuous surveillance and coverage of the coastline needs to start with ports, major cities and strategic islands. A combination of surface and underwater nodes, both fixed and mobile, could be built up into a capable defensive network. Machine Learning could be used to train this sensor net for highlighting suspicious activity, which could then be manually investigated. Security of ports and berths is a 24x7x365 task that deserves technology induction for automation as well as greater effectiveness against intruders.

Modern navies are investing in systems that provide persistent maritime surveillance. Unmanned underwater systems are being developed for ISR missions to provide persistent surveillance in contested maritime domains. For example, unmanned systems are used for monitoring coastal areas, gathering intelligence on enemy movements, and even interdicting illicit maritime activities. Such systems can operate in both littoral and deep-water environments, adapting their sensor payloads to the mission at hand.

Underwater Search and Anti-Submarine Warfare (ASW)

Searching for objects underwater, be it wreckage, moored or laid mines, or lurking submarines, is highly challenging. Manned submarines are highly prized assets with very limited numbers compared to their vast domain of operations. Unmanned vehicles and sensors could significantly enhance the effectiveness of submarines, if challenges of communication are addressed.

Conversely, a combination of fixed and mobile nodes/assets could enable coordinated search for a moving submarine. Anti-submarine warfare is one of the most critical roles where unmanned underwater systems could prove useful.

For example, the REMUS series, particularly the REMUS 600 and 6000 models, have reportedly been deployed as part of ASW missions. These vehicles employ high-frequency sonars and sophisticated inertial navigation systems. In future, such solutions could potentially be used to detect and monitor quiet diesel-electric submarines, providing early warning to anti-submarine forces.

Mine Countermeasures (MCM)

Unmanned systems are ideally suited for 'dull, dirty, dangerous' missions, and searching for mines is an apt example. AUVs can be used to detect suspicious objects, and these can be further investigated by ROVs and detonated remotely.

Vehicles like the REMUS series (Figure 1-9) have been widely adopted for MCM missions. These AUVs can autonomously map minefields and even carry out preliminary neutralization tasks.

Collaborative operation of an Unmanned Surface Vessel, carrying an AUV, has already demonstrated the ability to detect and neutralise mines remotely by a trained crew.[21] The ruggedisation and proliferation of such technology would greatly multiply the capability of navies in undertaking such tasks.

Swarm deployments are being tested to enhance mine countermeasure capabilities. In one notable trial, multiple small UUVs were coordinated to survey and neutralize minefields in a simulated contested environment. By working in concert, these vehicles were able to cover a larger area and provide redundant data streams, thereby reducing the likelihood of mission failure.[22]

Figure 1-9. REMUS 100 operated by a Norwegian Naval EOD Commando
(Source: Kongsberg[20])

---

[20]   Kongsberg, 2009

[21]   Thales, 2020

[22]   Cimino, 2023

Strike

Delivery of munitions or weapons using drones is perhaps their most demanding application. In experimental trials, prototype vehicles have been equipped with precision-guided munitions, allowing them to engage enemy vessels autonomously. Such capabilities, while still in the early stages, could fundamentally alter the balance of naval power in littoral combat scenarios.

The ongoing Russia-Ukraine conflict has provided several examples of weaponised USVs and AUVs used for offensive missions, sometimes in 'kamikaze' or suicide missions (Figure 1-10). This war has led to a re-appreciation of the capabilities and deployment of drones in warfare.[23] We can see video clips on our mobile phones of USV swarms attacking a surface ship in the dark by homing on to its infrared signature, and emerging successful despite countermeasures. Proliferation of such cost-effective and lethal unmanned platforms has the potential to overrun defences of warships through sheer numbers and self-destructing impact.[24]

**Figure 1-10. View from a USV on one-way mission
homing in on Russian warship**
(Source: Interfax Ukraine[25])

---

[23] Abdurasulov, 2024

[24] Dickinson, 2025

[25] Interfax Ukraine. August 4, 2023. https://www.marinelog.com/news/video-ukrainian-drone-inflicts-heavy-damage-on-russian-warship/

Recent deployments of aerial drones for both attack and defence of merchant shipping in the Red Sea are examples of rapidly diversifying use cases.[26] In the future, teaming of aerial, surface and underwater drones, as well as manned-unmanned collaboration, could provide unprecedented and flexible capabilities in complex maritime security scenarios.

Requirements for Defence Applications

Defence applications impose some of the most demanding requirements on unmanned underwater systems, which are linked to their applications or roles as indicated:

- **Stealth and Low Signature**: UUVs must operate covertly in contested environments, necessitating quiet propulsion systems and minimal acoustic signatures.

- **Autonomous ISR**: Platforms must be capable of gathering, processing, and transmitting critical information about enemy movements and underwater threats.

- **Mine Countermeasures (MCM)**: Effective detection, classification, and neutralization of underwater mines are paramount. This requires precise sensor integration and robust manipulator capabilities.

- **Rapid Deployment and Swarming**: Defence systems often need to be deployed quickly and in large numbers. Swarm intelligence and networked operations can overwhelm adversary defences.

- **Robust Communication and Data Security**: Secure, reliable communication links are vital in a contested electromagnetic environment, often requiring redundancy and low-latency protocols.

- **Manned-Unmanned Teaming**: Defence UUVs should be able to supplement manned systems, providing extended reach and persistence while reducing the risk to personnel.

To meet these challenges, focus areas for AUV technologies for defence applications are summarised in Table 1-2. These are in addition to the focus areas for technological solutions mentioned earlier (in Table 1-1) for scientific and industrial applications.

Table 1-2. Challenges and Solutions Required for Defence AUV Applications

Problem Area	*Challenges*	*Solutions Required*
Underwater Communication	Underwater communications are limited by the physics of acoustic transmission, which suffers from low bandwidth, latency, and susceptibility to interference.	Further research into multi-modal communication systems is necessary. Hybrid solutions that integrate acoustic, optical, and even limited RF-based methods may offer higher data rates and more robust connectivity. Enhanced encryption and adaptive routing protocols are also needed to ensure secure, reliable data exchange in contested environments.
Stealth and Low Acoustic Signature	Military UUVs must operate undetected in enemy waters. Noise from propulsion systems and electronic components can compromise stealth.	Continued development of quiet propulsion systems, such as advanced electric thrusters and biomimetic designs, is required. Research into noise-cancelling technologies and acoustic decoupling of onboard electronics can further reduce the acoustic footprint of these systems.
Autonomy in Contested Environments	The dynamic and hostile underwater battlefield requires UUVs to operate with a high degree of autonomy, making real-time decisions in rapidly evolving scenarios.	Advanced onboard AI and machine learning algorithms must be developed to handle complex threat assessments, obstacle avoidance, and coordinated swarm behaviour. Robust fail-safe mechanisms and redundancy in sensor systems will ensure operational continuity even when parts of the system are compromised.
Integration with Manned and Other Unmanned Systems	To be truly effective, unmanned systems need to be integrated into multi-domain network including manned ships, aircraft, and satellites.	Standardised communication protocols and data formats are essential for interoperability. Future systems should emphasize plug-and-play compatibility with legacy platforms as well as new unmanned systems, enabling seamless coordination and joint operations.

Cross-Domain Synergy

The applications of Unmanned Underwater Systems for scientific and industrial roles can also apply to defence applications. The tasks of environmental survey, underwater search, seabed mapping, and structural inspections are useful for maritime security applications and employ similar platforms and sensors.

Shared Technologies

While the scientific, industrial, and defence domains each impose unique requirements on unmanned underwater systems, there are significant areas of overlap that allow for cross-domain synergies. Advances in one area often benefit the others, and collaborative research can lead to more robust and versatile systems. Several common key technologies underpin applications across all domains:

- **Autonomy and AI**: Advanced algorithms developed for defence applications, where real-time decision making and adaptive responses are critical, can be adapted to enhance scientific surveys and industrial inspections. Similarly, the machine learning techniques used in processing complex sonar data for geological mapping have direct applications in mine detection and enemy tracking.

- **Sensor Integration**: High-resolution imaging, sonar mapping, and multi-sensor data fusion are universally required. Improvements in sensor miniaturization, energy efficiency, and data processing benefit scientific research, industrial monitoring, and military ISR alike.

- **Navigation and Localization**: Robust underwater navigation is a shared challenge. Hybrid systems that combine inertial navigation, acoustic positioning, and periodic GPS fixes are essential whether an AUV is mapping the seafloor for research or stealthily tracking enemy submarines.

- **Power Systems**: Enhancing energy density and developing innovative power management solutions are crucial for extending the operational range of UUVs in every domain. Industrial systems often push the limits of endurance, while defence applications require reliability and rapid redeployment.

- **Swarm operations**: By coordinating multiple UUVs to operate as a single, distributed sensor network, larger areas and several objects of interest can be monitored simultaneously. Such a coordinated approach could also be useful to overcome the limitations of mobility (speed) or endurance of small unmanned systems for all types of applications.

Looking Ahead

We have seen the basic features of uncrewed underwater systems in this chapter, including their major systems, components, and features. The deployment of unmanned underwater systems offers significant benefits across scientific, industrial and defence applications that have been described.

In scientific and industrial contexts, underwater unmanned systems significantly improve capabilities, operate beyond the limits possible with crewed operations, and minimise danger to human life. In defence, the ability to operate unmanned systems in high-risk environments provides a force-multiplying effect, while preserving valuable crews from avoidable risks.

The performance features required for various tasks in each of these three categories of application have been discussed. These were followed by the challenges (environmental, technical, operational) in achieving those performance features. Thereafter, the technology areas that need to be focused upon for finding solutions to those challenges have been tabulated.

We have seen the purposes and means of doing underwater surveys using UUVs. These highlight the commonality of features, requirements, and technological challenges for UUVs across various application domains. Synergy across sectors in investments and roadmaps for research and development is therefore essential to achieve the potential of underwater robotics.

The growing realisation of the significance of the oceans, the multi-disciplinary nature of the technologies involved, and close linkages with AI/ML developments, all make this field highly opportune and poised for rapid growth. In the next chapter we examine the developments in this field in India and other nations, and the roles of various stakeholders for enhancing the ecosystem to encourage their rapid development.

References

Abdurasulov, A. (2024, March 12). *Ukraine war: The sea drones keeping Russia's warships at bay.* Retrieved from BBC News: https://www.bbc.com/news/world-europe-68528761

Ballard, R. D., & Drew, C. (2021). *Into the Deep: A Memoir from the Man who Found Titanic.* National Geographic.

Butler, B. (2018). *Into the Labyrinth: The Making of a Modern-Day Theseus.* Canada.

Christ, R., Wernli, R., & Wernli Sr., R. (2011). *The ROV Manual: A User Guide for Observation Class Remotely Operated Vehicles.* Netherlands: Elsevier Science.

Cimino, V., Campagnaro, F., & Zorzi, M. (2023). A Mine Countermeasure System with

Low-Cost AUV Swarms. WUWNet 2023, Shenzhen, China. Retrieved from: https://signet.dei.unipd.it/wp-content/uploads/2023/11/vincenzo.pdf

Dunbabin, M., Roberts, Usher, J. K., Winstanley, G., Corke, P. (2005). A Hybrid AUV Design for Shallow Water Reef Navigation. Proceedings of the 2005 IEEE International Conference on Robotics and Automation, Barcelona, Spain. pp. 2105-2110, doi: 10.1109/ROBOT.2005.1570424.

DeChiaro, J. (2022, February 1). *What System Designers Should Know About MOSA Standards.* Retrieved from mobilityengineeringtech.com: https://www.mobilityengineeringtech.com/component/content/article/40871-what-system-designers-should-know-about-mosa-standards

Dickinson, P. (2025, June 12). *Ukraine is shaping the future of drone warfare at sea as well as on land.* Retrieved from AtlanticCouncil.org: https://www.atlanticcouncil.org/blogs/ukrainealert/ukraine-is-shaping-the-future-of-drone-warfare-at-sea-as-well-as-on-land/

Fedasiuk, R. (2021, August 17). *Leviathan Wakes: China's Growing Fleet of Autonomous Undersea Vehicles.* Retrieved from CIMSEC.org: https://cimsec.org/leviathan-wakes-chinas-growing-fleet-of-autonomous-undersea-vehicles/

Griffiths, G. (2003). *Technology and Applications of Autonomous Underwater Vehicles.* (G. Griffiths, Ed.) Taylor & Francis. doi:10.1201/9780203023249

Griffiths, G., Millard, N., & Rogers, R. (2003). Logistics, Risks and Procedures concerning Autonomous Underwater Vehicles. In *Technology and Applications of Autonomous Underwater Vehicles* (pp. 279-293). CRC Press. doi:10.1201/9780203522301.ch16

Hand, M. (2024, March 11). *28 Houthi drones shot down in Red Sea as attacks intensify.* Retrieved from Seatrade Maritime News: https://www.seatrade-maritime.com/casualty/28-houthi-drones-shot-down-red-sea-attacks-intensify

Kongsberg. (2009, June). *Naval AUV product range The HUGIN & REMUS Family.* Retrieved from www.kongsberg.com: https://www.kimerius.com/app/download/5785529677/The+Hugin++Remus+family.pdf

Pantheeradiyil, A. (2024, October 7). *How ROVs and divers are transforming underwater inspections for safety and efficiency.* Retrieved from blueyerobotics.com: https://www.blueyerobotics.com/blog/how-rovs-and-divers-are-transforming-underwater-inspections-for-safety-and

Ray, A., Singh, S., & Seshadri, V. (29 – 30 June, 2011). Underwater Gliders - Force Multipliers for Naval Roles. *Warship 2011: Naval Submarines and UUVs.* Bath, UK: Royal Institution of Naval Architects.

Real Time Innovations. (n.d.). *RTI Connext for the Unmanned Maritime Autonomy Architecture (UMAA): Using Modular Open Systems Architecture (MOSA) to Accelerate the Next Generation Navy.* Retrieved from rti.com: https://www.rti.com/hubfs/_Collateral/capability-briefs/rti_capability-brief-unmanned-maritime-autonomy-architecture.pdf

Rumson, A. G. (2021). The application of fully unmanned robotic systems for inspection of subsea pipelines. *Ocean Engineering,* 235. doi:https://doi.org/10.1016/j.oceaneng.2021.109214.

Subsea Pipeline Inspection. (2025, May). Retrieved from TSC Subsea: https://www.tscsubsea.com/pipeline-inspections/

Thales to deliver the World's First Fully Integrated Unmanned Mine Countermeasures System for the Royal Navy and French "Marine Nationale. (2020, November 26). Retrieved from: https://www.thalesgroup.com/en/group/journalist/ press_release/thales-deliver-worlds-first-fully-integrated-unmanned-mine

Unmanned Underwater Vehicles (UUV) Industry Research Report 2023-2030: Opportunities in Deep-Sea Exploration, Subsea Resource Mapping, Maritime Surveillance and Defense Applications. (2024). Retrieved from: https://www.globenewswire.com/news-release/2024/ 09/03/2939731 /28124/en/Unmanned-Underwater-Vehicles-UUV-Industry-Research-Report-2023-2030.html

Weir, G. E. (2001). *An Ocean in Common: American Naval Officers, Scientists, and the Ocean Environment.* Texas A&M University Press.

Xiang Liu, S. X. (2025). Innovative Strategy and Practice of Using Underwater Robot for Marine Cable Inspection and Operation and Maintenance. *Cognitive Robotics.* Retrieved from https://doi.org/10.1016/j.rineng.2021.100201

Canjun Yang, Xin Wu, Mingwei Lin, Ri Lin, Di Wu (2024). A review of advances in underwater humanoid robots for human–machine cooperation. *Robotics and Autonomous Systems,* 179. Retrieved from https://www.sciencedirect.com/science/article/abs/pii/S0921889024001283

Zhang, H. (2020). Subsea pipeline leak inspection by autonomous underwater vehicle. *Applied Ocean Research.* doi:10.1016/j.apor.2020.102321

2

Development Ecosystem

The first international conference on Unmanned Systems that I attended was at Jaipur in 2013. There were presentations on aerial, surface and marine robots: caterpillar-like motions, fish-like robots, and aerial vehicles like HALE and MALE that I had never heard before. When I presented my paper on 'Underwater Gliders', I found that it was quite novel for the audience, in spite of the wide range of engineering disciplines represented there. In an earlier conference in 2011 at Bath (UK) on Submarines and Underwater Technology, my paper on Underwater Gliders had also been novel for the specialised audience, mainly of Naval Architects. Even with common interests in technology areas, our individual disciplines can throw up such artificial barriers, with labels such as 'generalists' as well as 'specialists' getting swapped unpredictably.

The design and development of an engineering solution therefore requires not one inventor or one firm, but an entire ecosystem that connects user needs (both expressed and unspoken) with technological ability, means, trials facilities, infrastructure, and support for operations.

In this chapter we start with a chronology of UXV development in India. We then look at case studies of the development cycle for two of the most popular AUVs in the world, to draw lessons from their journey of product development, trials and induction. Timelines for development are compared and analysed. We then return to the Indian ecosystem of development agencies, and policy initiatives for underwater and marine robotics, before suggesting measures to develop this ecosystem further.

Indian Developments

The National Institute of Oceanography (of CSIR) at Goa was the first in India to develop an Autonomous Underwater Vehicle (AUV). This was *Maya*, developed in 2006, used for oceanographic data collection.[1] The project was funded by CSIR-NIO and the Ministry of Communications & IT, GoI (Figure 2-1).

Figure 2-1. CSIR-NIO Goa team with India's first AUV Maya
(Source: NIO)

In 2009-10, another CSIR lab, the Central Mechanical Engineering Research Institute (CMERI) at Durgapur, West Bengal, developed the AUV-150 along with IIT Kharagpur.[2] This project was funded by the Department of Ocean Development, which later became the Ministry of Earth Sciences (MoES). The AUV-150 was demonstrated at Idukki Lake (Kerala) and subsequently at sea in 2011.[3] A deeper-diving version, AUV-500, was also developed and tested off Goa.

Figure 2-2. CSIR-CMERI team with AUV-150
(Source: CMERI)[3]

---

[1]   Elgar Desa, 2007

[2]   Shome, et al., 2010

[3]   Shome, S.N. et al., 2010

Figure 2-3. Sea Trials of AUV-150 by CSIR-CMERI off Chennai in 2011[4]

The Naval Science and Technological Laboratory (NSTL) of the DRDO located at Visakhapatnam developed a 'flatfish'-form AUV around 2011 as a technology demonstrator.[5] This was also successfully tested at sea but was not inducted.

Figure 2-4. Sea Trials of Flatfish AUV by NSTL in 2011-12 (DRDO-NSTL)[5]

---

4   Agnihotri, 2023
5   Manu Korulla, 2016; https://www.drdo.gov.in/drdo/autonomous-underwater-vehicle-auv

The GoI's Department of Ocean Development (which became the Ministry of Earth Sciences (MoES) in 2006) set up the National Institute of Ocean Technology (NIOT) at Chennai in 1993. NIOT has developed ROVs (ROSUB 6000, PROVe),[6] which have been used for Indian Ocean as well as polar expeditions.

The technology of AUV *Maya* was transferred by NIO Goa to M/s Larsen and Toubro,[7] who subsequently demonstrated (from 2014) a series of AUV designs including *Adamya* and *Amogh*, developed with foreign collaboration.[8]

Recent Indian Developments

In spite of such pioneering developments since 2006 by CSIR-NIO, CSIR-CMERI and DRDO-NSTL, as well as their technology transfer to industry, these AUV designs did not translate to production orders at scale.

In the past decade, the possibilities and applications of aerial drones have captured public imagination. This interest is benefiting marine robotics as well. With the growing capability of autonomous systems (computing power, compact electronics and affordable components), there has been a noticeable increase in interest in marine and underwater robotics.[9]

Since about 2020, a large number of enthusiastic start-ups across the country have become engaged and gained proficiency in AUV / ROV development. These include M/s Planys, TSC Tech, Arbotnix, Sagar Defence, Rekise Marine, EyeROV, Hyperhorizon and several others. Large Indian firms engaged in the development of UXVs include M/s Aerospace Engineering, Larsen & Toubro, and Kalyani Strategic Systems (Bharat Forge). Shipyards such as GRSE, MDL and CSL have also got involved in design/development/ production of UXVs in various capacities such as incubating, sponsoring, funding, collaborating, fabricating and trials.

Examples of ROVs, ASVs and AUVs developed in India since 2015 by various institutions and firms are shown in Figures 2-5, 2-6, 2-7 and 2-8.

---

[6] Sathianarayanan, 2013; https://www.niot.res.in/niot_deepseatech_en.php

[7] https://www.nio.res.in/research/technologies/autonomous-underwater-vehicles-auv

[8] Kullashri, 2022

[9] Agnihotri, 2023

Jalasimha by Coratia Tech, Rourkela[10]	*Tuna* by EyeROV, Kochi[11]
Beluga by Planys Technologies, Chennai[12]	*ROV500* by CSIR-CMERI[13]

Figure 2-5. Examples of ROVs developed in India

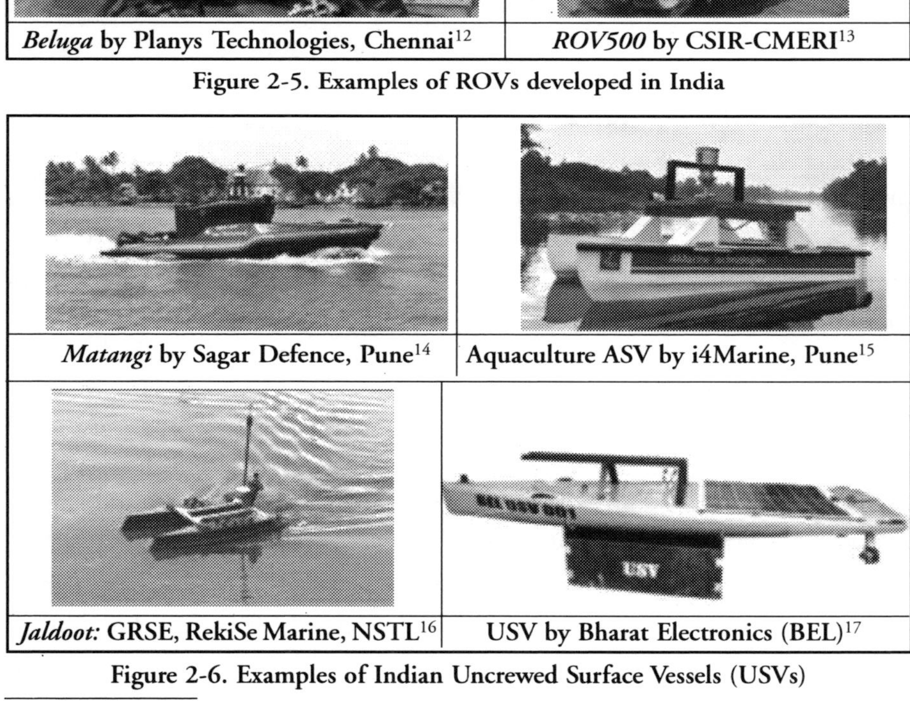

Matangi by Sagar Defence, Pune[14]	Aquaculture ASV by i4Marine, Pune[15]
Jaldoot: GRSE, RekiSe Marine, NSTL[16]	USV by Bharat Electronics (BEL)[17]

Figure 2-6. Examples of Indian Uncrewed Surface Vessels (USVs)

---

[10] https://coratia.com/product/2
[11] https://eyerov.com/products/tuna/
[12] https://planystech.com/technology/
[13] https://www.cmeri.res.in/technology/remotely-operated-vehicle-rov
[14] https://x.com/indiannavy/status/1851593148620796287; https://www.sagardefence.com/;
(Tyagi, 2024)
[15] https://www.i4-marine.com/
[16] Business Line, Dec 2024
[17] BEL's Zero Emission Unmanned Surface Vehicle, 2024

HE-AUV by DRDO and CSL, Kochi[18]	*Amogh* by L&T and Edgelab[19]
Neerakshi by GRSE and AEPL[20]	*Svaayatt* by Planys Technologies[21]

Figure 2-7. Examples of AUVs developed in India

Coral-Bot by CSIR-NIO, Goa[22]	*Sunplower* by Aritra, Hyderabad[23]

Figure 2-8. Examples of Indian AUVs with unconventional forms

---

[18]   https://x.com/cslcochin/status/1765917170427670683

[19]   https://www.lntpes.com/our-offerings/marine-platforms-equipment-and-systems/autonomous-naval-platforms/

[20]   https://www.hindustantimes.com/india-news/india-launches-neerakshi-autonomous-underwater-vehicle-for-mine-detection-101690636002474.html; Singh, 2023

[21]   https://planystech.com/technology/

[22]   https://www.oceansociety.in/docs/ocean/ocean_digest/2023/Ocean-Digest_2023%28Vol_10%29_Issue_3.pdf ; Maurya, 2023; Souza, 2024

[23]   https://www.sunplower.com/marine

Highly capable technology start-ups such as M/s Xalten, Yaanendriya, Hydrovert, i4Marine, Nautical Wings, Raphe MPhibr, Airbotix, etc., as well as industry stalwarts such as M/s Keltron and M/s Mahindra have been developing components, technologies and capabilities that have great potential for underwater robotics applications. Several firms possess expertise in inducting niche technologies related to marine robotics, such as M/s Samhitha Marine, Norbit, Pan India, etc.

Despite advances, significant technological challenges persist, such as limited battery endurance, few indigenous sensors, and communication constraints. Integration of complex subsystems remains problematic, impacting operational robustness and mission success rates. Paucity of skilled personnel and limited interdisciplinary coordination slow down progress. Financial constraints and limited commercialization pathways reduce incentives for scaling up production. Dependence on imported critical components undermines self-reliance goals. The underwater environment's unpredictability, and long gestation period for research, further complicate new development. Additionally, changes in regulatory and policy environments typically lag behind technological innovation, affecting commercialization and international collaboration.

The perceived limitations in maturity of marine robotics in India are represented in Figure 2-9.

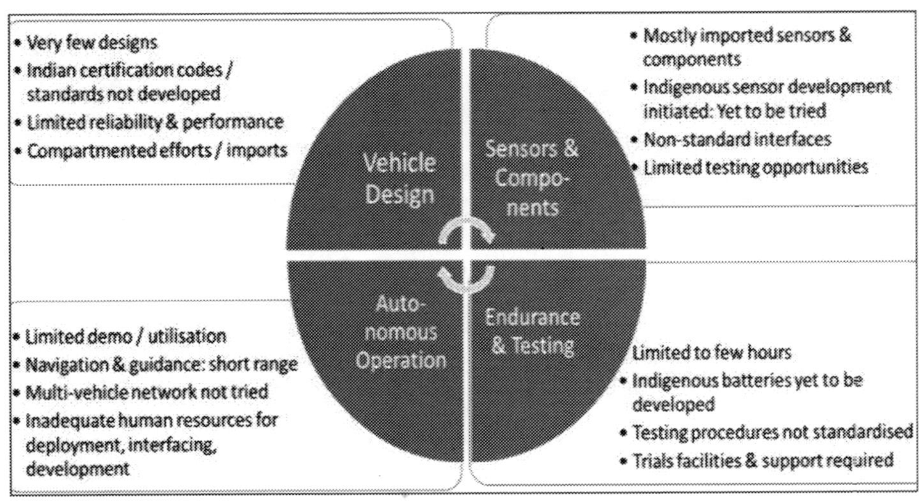

Figure 2-9. Author's Assessment of Maturity of Marine Robotics in India

Case Study: *Hugin* AUV Development

The Norwegian conglomerate Kongsberg manufactures the HUGIN series of AUVs, besides providing various instruments and software for navigation, as well a plethora of other technologies. The HUGIN AUV can be used for subsea sensing up to 6,000 metres depth. The design and development cycle of the HUGIN series is discussed further in this case study.

Overview and Background

In the 1980s, Kongsberg developed a small AUV for technology and test purposes. In 1995, the HUGIN project was created in collaboration with Statoil, Norwegian Defence Research Establishment (FFI), Norsk Undervannsintervensjon (NUI) and Kongsberg. Since its inception in the early 1990s, the HUGIN AUV family (Figure 2-10) has evolved into what is widely regarded as the most successful commercial AUV platform available.

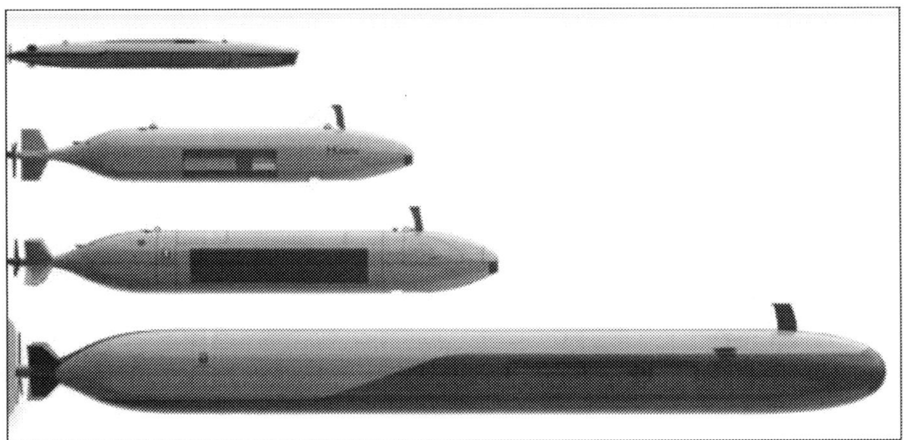

Figure 2-10. The HUGIN range of AUVs (Top to bottom: HUGIN Edge, HUGIN, HUGIN Superior and HUGIN Endurance)
(Source: Kongsberg[24])

The programme's first prototype made its inaugural dive in 1993, and the first commercial survey was conducted in 1997. Over the past three decades, continuous innovation has allowed the HUGIN series to meet increasingly demanding applications in hydrographic surveying, oil and gas exploration, scientific research, and defence operations.[25]

---

[24] https://www.kongsberg.com/discovery/autonomous-and-uncrewed-solutions/auv/
[25] Kongsberg, 2009

Concept and Innovation

- **Early R&D and Partnerships.** The HUGIN program was born from the need to develop an un-tethered underwater survey platform. Early concepts emphasized autonomous operation with integrated navigation systems. Over time, innovations such as plug and play payload systems and advanced battery technology were introduced to enhance versatility and endurance.

- **Modular Architecture and Sensor Integration.** The design evolved to incorporate a robust aided inertial navigation system (INS) fused with Doppler velocity log (DVL) measurements, GPS surface fixes, and—where possible—terrain-aided navigation updates. This 'toolbox' of aiding techniques enables the vehicle to maintain high positioning accuracy even during extended underwater operations.[26]

- **Evolution of Models.** Early HUGIN models focused on meeting the basic requirements for detailed seabed mapping. Later models such as HUGIN Endurance and HUGIN Edge were developed with additional features.

  o **HUGIN:** Since introduction in the 1990s, HUGIN AUVs have completed more commercial surveys than any other AUV. Its length is from 5.2 to 6.4 metres, weighing 1 to 1.5 tons. The model is available for operating at depths of 3,000, 4,500, and 6,000 metres.

  o **HUGIN Superior:** It has a greater length of 6.6 metres, weighing about 2.2 tons, with a depth rating of 6,000 metres and improved payload sensors.

  o **HUGIN Edge:** A compact, sub 4 metre AUV (approximately 300/ kg) optimized for rapid deployment from unmanned surface vessels (USVs) or from shore, featuring over 24 hours of endurance at depths up to 1,000 metres.

  o **HUGIN Endurance:** An 8 ton, 12-metre vehicle designed for long-range, fully autonomous missions with a reported operational range of 1,200 nautical miles (2,200 km) and up to 15 days at sea.

---

[26] Jalving, Gade, Hagen, & Vestgård, 2004

Development Methodology

- **Iterative Engineering Process.** The design cycle for HUGIN AUVs is characterized by a continuous loop of concept validation, prototype development, sea trial testing, and software/hardware updates. Early design decisions, ranging from hull shape and battery configuration to payload integration, were revisited iteratively as real-world data from field trials became available.

- **Integration of Advanced Navigation and Autonomy.** Subsequent generations have benefited from developments in terrain navigation algorithms (often based on point mass filter estimators) that allow the vehicle to restrict navigation drift using onboard multi-beam echo sounders and pre-loaded digital bathymetric maps. This integration has been the key to the vehicle's performance during prolonged autonomous missions.

Test Protocols and Trials

Tests and trials served two major purposes:

- **Redundancy and Integrity Testing.** The trials included a series of integrity tests to ensure that only position fixes meeting strict convergence and goodness-of-fit criteria were passed to the vehicle's primary navigation system. This conservative approach (even if it resulted in a high rejection rate of intermediate fixes) has been vital in maintaining overall system accuracy.

- **Iterative Feedback to Design.** Data from these trials informed subsequent design modifications—including updates to the autonomy framework and sensor integration—which have continuously improved both reliability and operational performance.

Earlier models (e.g., HUGIN 1000) underwent rigorous factory acceptance tests and sea trials. These trials, reported in 2002, included scenarios where aiding sensors (such as the DVL and terrain navigation modules) were selectively disabled to demonstrate the system's resilience.[27] In another documented trial, a HUGIN 1000 vehicle maintained navigation accuracy within 5 metres over a 7 hour transit in challenging environments.[28]

---

[27] Jalving, Vestgård, & Storkersen, 2002
[28] Hagen, Anonsen, & Mandt, 2010

One of the notable trials was a recent record-setting mission of the 'HUGIN Endurance'. In a fully autonomous sortie conducted between September 2023 and March 2024, the vehicle operated without any human intervention or external navigation support.[29] Key trial results included the following:[30]

- **Depth Range:** The vehicle operated at depths ranging from 50 metres to 3,400 metres.

- **Mission Range and Duration:** It covered 1,200 nautical miles during the multi-week mission.

- **Survey Capability:** The mission included imaging and bathymetry over 36 square nautical miles using synthetic aperture sonar (SAS) and laser profiling.

- **Navigation Accuracy:** Despite the absence of continuous external fixes, the vehicle returned with a position error of approximately 0.02 per cent of the total distance travelled.

Induction Cycle and Market Adoption

After successful sea trials and rigorous internal testing, HUGIN AUV models undergo a formal induction cycle that includes certification processes, factory acceptance tests, and integration into customer platforms (whether from dedicated survey vessels, USVs, or shore-based systems).

Customers are introduced to the systems through comprehensive operator training and support programs. The HUGIN system is designed for 'plug-and-play' integration, enabling rapid deployment in diverse operational scenarios.

Since the early days of the product line, more than 100 HUGIN AUV systems have been delivered worldwide. Today, the platform is used by both commercial operators (in oil and gas, renewables, hydrographic surveying) and government customers.

Notably, over 12 navies currently operate HUGIN AUVs, for tasks ranging from intelligence preparation of the operational environment (IPoE) to mine countermeasures (MCM) and subsea warfare. Recent contracts include a Defence Innovation Unit (DIU) award for the US Navy's large displacement UUV program.[31]

---

[29] HUGIN Endurance AUV Smashes Records in Multi-week Fully Autonomous Mission, 2024

[30] Kongsberg, 2024

[31] Kongsberg Discovery Wins US Navy DIU Contract, 2024

In the second quarter of 2022, Kongsberg Maritime's Sensor and Robotics division secured orders valued at over NOK/ 450 million (approximately € 43.7 million), while another report indicated contracts totalling around US$ 44 million during the same period.[32] These orders represent both recurring business from established customers and new contracts from emerging markets.

Summary and Future Outlook

The HUGIN AUV case study illustrates a comprehensive and iterative process—from concept to design, rigorous trials, and finally, successful customer induction. Key elements include:

- **Innovative Modular Design.** Continuous improvements in hull design, battery technology, and sensor integration have allowed the platform to remain at the cutting edge of underwater robotics.

- **Robust Navigation and Autonomy.** Advanced aiding techniques (combining INS, DVL, and terrain navigation) ensure that even during extended autonomous operations, the vehicle maintains high positional accuracy.

- **Successful Sea Trials and Market Penetration.** Field trials—such as the multi-week HUGIN Endurance mission—demonstrate the vehicle's operational capability under extreme conditions. Subsequent successful induction into diverse operational roles is evidenced by strong contract awards and an expanding global customer base.

- **Growing Market Acceptance.** With over 100 systems delivered globally and significant orders secured in recent years, the HUGIN family continues to set industry standards. The platform's adaptability—from compact models like HUGIN Edge to long-range systems like HUGIN Endurance—positions it well for future expansion in both commercial and defence markets.

As the technology advances further (with ongoing software updates and potential enhancements in energy and sensor technology), HUGIN AUVs are expected to consolidate their market leadership and continue to support critical underwater operations worldwide.

---

[32] Skopljak, 2022

Case Study: *Bluefin* AUV

Bluefin Robotics, founded in 1997 by a core group of engineers from the MIT Autonomous Underwater Vehicle Lab, is the maker of the Bluefin-21 AUV.

Overview and Design Objectives

The Bluefin-21 (Figure 2-11) was conceived as a compact, highly modular underwater platform intended for a wide variety of missions. Key design objectives included:

- **High modularity.** To allow rapid payload exchange and in-field maintenance.

- **Robust navigation.** Achieving high accuracy through the integration of inertial navigation systems (INS), Doppler velocity logs (DVLs), and, when possible, surface GPS fixes.

- **Efficient propulsion and power.** To provide a balance between speed and endurance, powered by an array of pressure-tolerant lithium-polymer battery packs.

- **Compact size and transportability.** With a torpedo-shaped hull of approximately 5.0 metres in length and 0.5 metres in diameter, the system is designed for ease of launch and recovery from relatively small vessels, including submarine torpedo tubes (that have a typical calibre of 533 mm).

The design philosophy was driven by the need to provide a versatile survey and reconnaissance tool that could operate independently for extended periods (approximately 25 hours at a cruising speed of 3 knots) and be reconfigured rapidly to meet different mission requirements.[33]

Concept and Preliminary Design

- **Mission Requirements Identification.** Early in the design process, Bluefin Robotics (and its early commercial partners) defined the key mission areas: offshore survey, search and salvage, underwater archaeology, oceanography, and mine countermeasures.

- **Modularity as a Key Driver.** The system was conceptualized with interchangeable payload modules (e.g., side-scan sonar, sub-bottom profilers, and multi-beam echo-sounders) and swappable battery

---

[33]   Lundquist, 2023

Figure 2-11. Bluefin-21 AUV being lowered for operation
(Source: General Dynamics[34])

packs. This approach was intended not only to reduce downtime between missions but to also allow the platform to be tailored quickly to different operational scenarios.

- **Hydrodynamic Considerations.** The design team adopted a torpedo-shaped hull—a common solution for underwater vehicles—to minimize drag and optimize endurance. Early computational fluid dynamics (CFD) simulations and empirical model tests helped fine tune the shape. For example, early blueprints and bench-scale prototypes were evaluated using towing tank tests and CFD tools to verify that the hull form would meet the desired performance at low to moderate speeds (approximately 3 to 4.5 knots).

Detailed Design and Engineering

- **Subsystem Integration.** Detailed engineering work focused on integrating the modular payload bay with the battery compartment and propulsion system. The ducted thrusters, mounted on a gimballed platform, were engineered to ensure that the vehicle could achieve its maximum speed (up to 4.5 knots) without compromising manoeuvrability.

---

34 https://gdmissionsystems.com/products/underwater-vehicles/bluefin-21-autonomous-underwater-vehicle

- **Navigation and Communication.** The navigation system was designed to achieve a dead-reckoning drift of less than 0.1 per cent of distance travelled. This was accomplished by integrating an INS with a DVL and supplementing these with periodic GPS fixes when the vehicle surfaced. An ultra-short baseline (USBL) acoustic positioning system was included for in-mission position updates.

- **Power System and Endurance.** Nine pressure-tolerant lithium-polymer battery packs, each with a rated energy of about 1.5 kWh, were arranged to meet the endurance requirement of 25 hours at 3 knots. Power management strategies were implemented in the vehicle's control software to optimize energy consumption during high-demand phases (e.g., during rapid transit) versus low-speed survey modes.

- **Control Software.** A robust control architecture was developed to handle the autonomous navigation, mission execution, and fault detection. The operator's tool suite, running on a Windows-based platform, allowed for mission planning, real-time monitoring, and post-mission data retrieval.

Trials and Testing

The Bluefin-21 underwent several phases of testing before being inducted into operational service:

- **Laboratory Testing.** Individual subsystems—such as the thruster, battery packs, sensor arrays, and communication modules—were initially tested in controlled laboratory environments to validate performance against specifications.

- **Towing Tank and Shakedown Tests.** Early prototype models were evaluated in towing tanks to measure hydrodynamic drag, stability, and manoeuvrability. These tests helped refine the hull design and control algorithms.

- **Field Trials.** The fully assembled Bluefin-21 participated in extensive sea trials. One notable field trial was during the US Navy's biennial Ice Exercise (ICEX) where a Bluefin-21 (named 'Macrura') was tested under Arctic ice conditions.[35] These trials verified the following capabilities:

---

[35]   ICEX 2020: U.S. Navy Deploys Autonomous Bluefin-21 UUV Under The Ice, 2021

    o Autonomous navigation under challenging environmental conditions.

    o Effective operation of modular payloads (such as side-scan sonar) during survey missions.

    o Robust communication protocols (acoustic and satellite) during the mission.

- **Mission Demonstrations.** The Bluefin-21 was later deployed in high-profile operational scenarios, including participation in the search for Malaysia Airlines Flight MH370.[36] These missions validated its ability to operate independently, conduct detailed seabed surveys, and reliably store and transmit large volumes of data. They provided crucial operational feedback that further improved the system's robustness and re-configurability for subsequent missions.

Induction and Operational Service

After iterative improvements from field trials, the Bluefin-21 was certified by both commercial clients and defence agencies. In the Final Acceptance Testing, its performance metrics (endurance, navigation accuracy, payload interchangeability, and safety features) met or exceeded the original design specifications.

The Bluefin-21 entered service in the early 2010s, with commercial operators (such as Phoenix International) and defence agencies (including the US Navy and other research organizations) adopting the system for routine missions.

Even after induction, Bluefin Robotics has continued to release firmware and hardware updates, demonstrating a 'live' induction cycle where lessons learned from each mission feed back into design improvements.

Key Lessons and Success Factors

- **Modular Design.** The modular payload and battery architecture has allowed the Bluefin-21 to flexibly adapt to various mission needs without a complete redesign.

- **Rigorous Testing.** A phased testing regime from lab and towing tank tests to extensive field trials proved essential to refining the AUV's performance.

---

[36] Taylor, 2014

- **Efficient Integration of Subsystems:** Early integration of propulsion, power, and navigation systems into a coherent whole minimized unforeseen issues during later field trials.
- **Feedback Loop.** The Bluefin-21's involvement in real-world missions provided invaluable feedback from operational use, which continues to inform incremental improvements and system upgrades.

UUV Development Timelines

The approximate development timelines (from concept to operational induction) for a range of contemporary AUV/UUV systems are summarised in Table 2-1. These durations are based on publicly available information and may vary depending on project scope, funding, technological challenges, and operational requirements.

Table 2-1. Development Timelines for Various AUVs

AUV/UUV Name	Manufacturer/ Country	Development Start	Prototype Trials	Operational Induction	Approx. Total Duration
Bluefin-21	Bluefin Robotics (USA)	~2008–2009	2012–2014 (field trials, including ice trials)	Early 2010s (by 2014–2015)	~5–7 years
REMUS 6000	WHOI/ Hydroid (USA)	Late 1990s	Early 2000s (multiple iterations and tow tank tests)	Mid-2000s	~5–8 years
Echo Voyager/ Orca XLUUV	Boeing/USA (with HII)	~2014–2016	2017–2022 (iterative sea trials)	2023 (first delivery)	~7–9 years
Manta Ray	Northrop Grumman/ USA (DARPA)	~2020	2022–2024 (at-sea trials)	2024–2025 (anticipated production phase)	~4–5 years
Qilin ARV	CSSC/China	2016	2019–2021 (prototype trials)	Late 2021 (inducted)	~5 years
Ghost Shark XL-AUV	Anduril Industries/ Australia	2022 (contract signed)	2023–2024 (prototype development and trials)	Mid-2025 (planned induction)	~3 years
Poseidon 6000 AUV	CSIC/China (Poseidon series)	Early 2020s (concept phase)	Ongoing prototype testing (varies by model)	Expected 2025 (for deep diving models)	~3–5 years (est.)

The table shows that while development timelines can vary widely depending on the system's size, intended mission, and technological complexity, many modern AUVs typically require between 3 to 9 years from concept to induction. Specific features for various types are as follows:

- **Bluefin-21**. Its timeline reflects an evolution from concept, through field trials, to induction in commercial and defence operations.

- **REMUS 6000**. A product of iterative research in academic and defence contexts, its development spanned from early experiments to a mature commercial product.

- **Echo Voyager/Orca XLUUV**. These platforms show a longer cycle due to the complexities of large, deep-diving unmanned systems intended for critical naval operations.

- **Manta Ray**. This represents a more recent effort focused on modularity and long-range, autonomous capability for military applications.

- **Qilin ARV**. A Chinese hybrid system combining autonomous and remotely controlled modes; its development was relatively rapid.

- **Ghost Shark**. An example of an ultra-fast-track development cycle aimed at fulfilling emerging defence requirements with a focus on stealth and modularity.

- **Poseidon 6000 AUV**. One of the latest in a series of Chinese AUVs designed for deep-sea exploration, with timelines still under refinement.

Trends such as rapid prototype testing (as seen with Ghost Shark) and hybrid operational modes (as with Qilin ARV) are helping to shorten these cycles and meet the growing demand for autonomous undersea capabilities in both defence and commercial sectors.

By analyzing these case studies and timelines, defence planners and manufacturers can better understand the strategic, technical, and operational considerations required to successfully design and field the next generation of unmanned undersea vehicles.

Indian Policy Initiatives

The impetus on indigenisation and innovation has resulted in multiple naval agencies being engaged today in this exciting field. The Indian defence start-up ecosystem has seen a growth spurt in the past three years as a result of the innovation organisations set up by the Ministry of Defence (MoD) and the armed forces' service headquarters. The Department of Defence Production's DIO/iDEX initiative[37] has been leveraged by the Indian Navy and more than 75 technology challenges were completed through Indian industry in a single year (2022-23).[38] Several of these have successfully transitioned into production orders for novel systems, which include marine robotics applications and solutions. These efforts in the naval domain are covered further in the next chapter.

Apart from the Navy's roadmap, there are two ongoing national missions that are highly relevant for marine robotics. These initiatives are summarised in Figure 2-12.

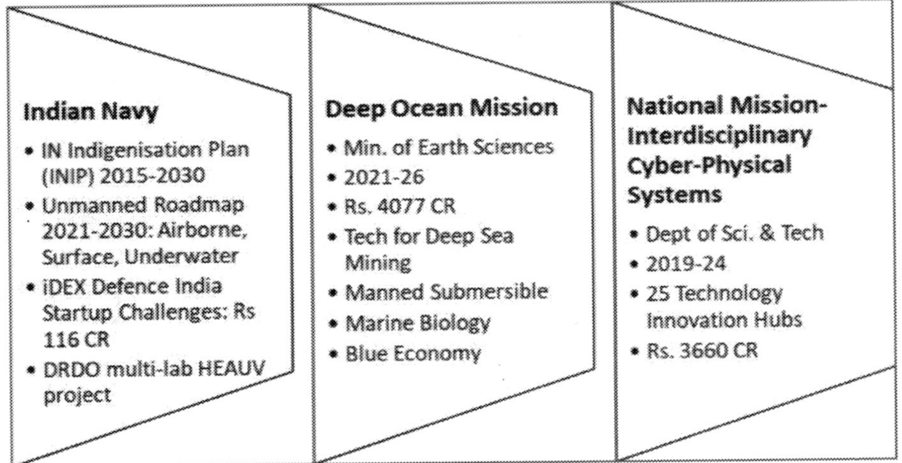

Figure 2-12. Major ongoing national missions relevant to marine robotics

The Deep Ocean Mission was approved by the Central government in 2022, with a sanction of Rs. 4,077 crore over a five-year period.[39] The Ministry of Earth Sciences is steering the Mission, which includes development of a deep ocean manned submersible by NIOT Chennai.[40] The mission also includes development of AUVs and related research projects.

---

[37]　https://idex.gov.in/challenge-categories

[38]　https://mod.gov.in/sites/default/files/PR041023.pdf, 04 Oct 23

[39]　https://moes.gov.in/schemes/dom?language_content_entity=en

[40]　https://www.moes.gov.in/programmes/ocean-technology/manned-and-unmanned-underwater-vehicles

The National Mission for Interdisciplinary Cyber-Physical Systems (NM-ICPS) steered by the Department of Science and Technology (DST) is another ambitious programme started in 2019.[41] Twenty-five 'Technology Innovation Hubs' (TIH) have been set up in partnership with reputed educational institutions across the country, each focusing on certain aspects of Cyber-Physical Systems. These not-for-profit companies function as start-up incubators (with the objective of becoming self-sustaining) and also promote technology development involving academia and industry.

Various government ministries and agencies related to development and applications of Marine Robotics in India are listed in Figure 2-13.

Developing the Ecosystem

Applications in the marine domain have less visibility than the multiplicity of aerial drones over land. The small size of the market leads to few takers in industry, and hence the negative cycle of low demand for indigenous specialised products. Articulation of 'blue-sky' needs is typically not a forte of the user. Demonstration of successful products is usually the best way to attract users, who can then envision further applications, which, in turn, drive production and technology development.

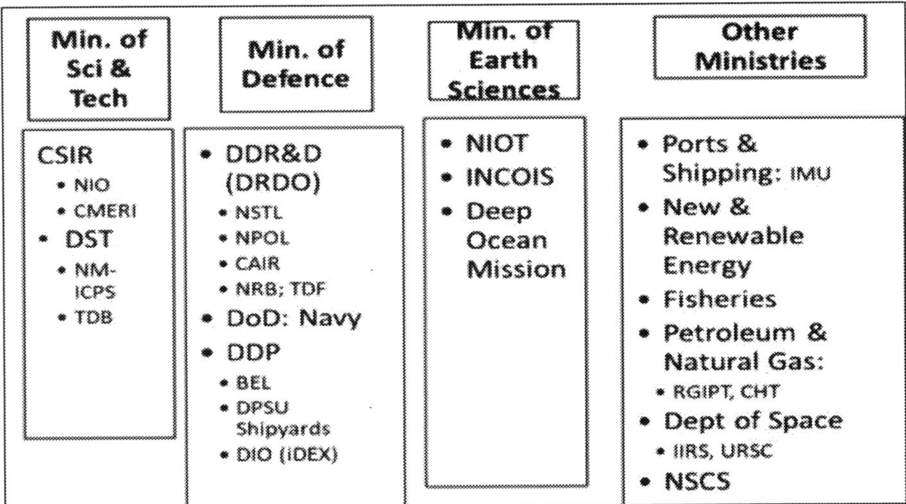

Figure 2-13. Government agencies linked to marine robotics and its applications

---

[41] https://nmicps.in/NM-ICPS-Scope

Objectives and Stakeholders

The measures proposed for developing national expertise in marine robotics would include the following:

- Gaining and developing expertise (knowledge and skills)
- Encourage R&D through funding
- Utilisation of robotics solutions by users, boosting demand
- Trials: Facilities, competence and experience
- Scaling up components and solutions
- Standards for production, evaluation and inter-operability
- Facilities for development and testing

There is a need to synergise efforts so that these required components of institutions, knowledge, skills, facilities and a healthy market can be developed for marine robotics. The activities and envisaged stakeholders in the 'ecosystem' of marine robotics are shown in Table 2-2.

Table 2-2. Components of the Marine Robotics Ecosystem

Component	Linked Activities	Lead Agency
Institutions	• Funds and facilities • Employ and nurture	Government
Knowledge	• Learning by sharing • Updates on progress • Innovative applications	Academia
Skills	• Students' outreach • Upskilling and cross-training	R&D labs
Facilities	• Integration and testing • Trials and evaluation	Users
Market	• Demonstrate and publicise • Component manufacturing • Connect with users	Industry

Technology Focus Areas

Specific areas of technology that are considered relevant for R&D in this field are as follows:

- Coordinated (swarm) operations
- Autonomous behaviour

- Underwater communications
- Shore control stations
- Launch and recovery systems
- Integration with satellite networks
- Indigenous sensors and components
- Intelligent data processing
- Energy storage and replenishment
- Standards for design and certification
- Materials development

Proposals for Indian Ecosystem

The enabling environment for innovation by tech start-ups needs to be complemented and backed by quick orders from users, as well as sustained production capabilities by industry. The R&D efforts for breakthrough technologies need to be funded and driven increasingly by the private sector, not just by government. A climate of collaboration and easing procedural hurdles must become the national endeavour at every level of governance.

Some specific proposals for encouraging and sustaining indigenous marine robotics capability are listed in Table 2-3.

Looking Ahead

The youthful human resource of the country, with the current buzz about AI and the opportunities of the oceans, hold tremendous promise for marine robotics. Unmanned and autonomous systems require development of innovative vehicles and components, ranging from as small as a few kilograms, useful not only in the oceans but also for any water body on land. In the years ahead, mankind would increasingly look to the oceans for food, energy and commerce. Unmanned and autonomous systems will play a significant role in these endeavours.

The Indian ecosystem for marine robotics is not just a necessity but an imperative for strategic indigenous capability. The global market in marine unmanned systems also offers export opportunities to Indian industry. To meet these aspirations, all stakeholders and agencies need to synergise efforts and collaborate to build a robust, indigenous, and self-sustaining ecosystem.

Table 2-3. Proposals for Enhancing Indian Marine Robotics Ecosystem

Area	*Recommendations*	*Agencies*
Institutions	• Empowered inter-ministerial Expert Group with overall perspective for coordinating and pooling resources across ongoing national missions • Distributed 'Centre of Excellence' in Marine Robotics, with nodes at various institutions • Collaboration through exchange of knowledge, personnel and facilities (rather than funds)	National Missions, Research Parks, Incubation Centres, Technology Innovation Hubs, User agencies
Knowledge	• Courses in robotics • Student participation in races/challenges • Internships with industry partners and R&D labs • Cross-discipline exposure • Development of standards	Academic institutions, Users, Industry
Skills	• Marine Robotics School (NIO Goa) could be an annual collaborative event • Annual workshops, funded by institutions	R&D agencies, industry, Users
Facilities	• Integration & Testing Facility near waterfront, accessible to all stakeholders • Designated ocean trials area for scientific studies, with connectivity • Instrumented trials facility for AUVs	Users (Navy, ONGC, etc), National Missions
Market	• Internal investments and R&D • Technology challenges (such as iDEX) with user interactions • Component manufacturing initiatives; More jobs in manufacturing sector • Create opportunities for fresh engineers in UXV design, R&D, testing, maintenance and operations	Industry (Shipyards), Users (Navy, Ministries of Surface Transport, Fisheries, Petroleum & Natural Gas, etc.)

Establishing a virtual or distributed Centre of Excellence incorporating government, academia, and industry would foster knowledge sharing, integrated R&D planning, and skills development. An inter-ministry framework for coordinating national efforts in the field of Underwater Robotics would also be highly beneficial.

In the next chapter we review the evolution of UUVs for naval applications, roadmaps and developments in various countries.

References

Agnihotri, K.K. (2023, September). Naval Drones: Force Multipliers in Naval Operations. *Synergy: Jl. of the Centre for Joint Warfare Studies, 2* (2) (pp.222-243). Retrieved from: https://cenjows.in/wp-content/uploads/2023/10/Synergy-Journal-online-version-merged.pdf

BEL's Zero Emission Unmanned Surface Vehicle (USV). (2024, April 12). *Bots & Drones India.*

BL New Delhi Bureau. (2024, December 5). GRSE hands over unmanned surface vessel 'Jaldoot' to DRDO. *The Hindu Business Line.*

Elgar Desa, R. M. (2007). The small AUV Maya: Initial Field Results. *International Ocean Systems, 11* (3).

Hagen, O., Anonsen, K., & Mandt, M. (2010). The HUGIN real-time terrain navigation system. *OCEANS 2010* (pp. 1-7). Sydney: IEEE.

HUGIN Endurance AUV Smashes Records in Multi-week Fully Autonomous Mission. (2024, September 4). Retrieved May 2025, from Marine Technology: https://www.marinetechnologynews.com/news/hugin-endurance-smashes-records-639764

Hydroid Inc. (2012, July). *Marine Technology Reporter.* Retrieved June 2025, from www.seadiscovery.com: https://magazines.marinelink.com/Magazines/MarineTechnology/201207/page/56

ICEX 2020: U.S. Navy Deploys Autonomous Bluefin-21 UUV Under The Ice. (2021, April 6). Retrieved May 2025, from General Dynamics Mission Systems: https://gdmissionsystems.com/articles/2021/04/26/in-the-news-us-navy-deploys-autonomous-uuvs-under-the-ice

Jalving, B., Gade, K., Hagen, O., & Vestgård, K. (2004). A Toolbox of Aiding Techniques for the HUGIN AUV Integrated Inertial Navigation System. *Modeling, Identification and Control.*

Jalving, B., Vestgård, K., & Storkersen, N. (2002). Detailed Seabed Surveys with AUVs. In G. (. Griffiths, *Technology and Applications of Autonomous Underwater Vehicles.* CRC Press.

Kirchhoff, C., & Shah, R. M. (2024). *Unit X: How the Pentagon and Silicon Valley Are Transforming the Future of War.* Scribner.

Kongsberg Discovery Wins US Navy DIU Contract. (2024, February 9). Retrieved May 2025, from MarineTechnologyNews.com: https://www.marinetechnologynews.com/news/kongsberg-discovery-contract-634492

Kongsberg. (2009, June). *Naval AUV product range The HUGIN & REMUS Family.* Retrieved from www.kongsberg.com: https://www.kimerius.com/app/download/5785529677/The+Hugin++Remus+family.pdf

Kongsberg. (2024, September). *Smashing AUV Records - HUGIN Endurance Completes a Multi-Week Fullly Autonomous Mission.* Retrieved May 2025, from kongsberg.com: https://www.kongsberg.com/newsroom/news-archive/2024/smashing-auv-records-hugin-endurance-completes-a-multi-week-fully-autonomous-mission/

Korulla, M. (April 2016). Autonomous Underwater Vehicles: Roadmap for a Future Ready Naval Force. *International Seminar on 'Make in India' Paradigm - Roadmap for a Future Ready Naval Force.* New Delhi: FICCI.

Kullashri, V. (2022, December 3). *Unmanned Underwater Vehicle – The Invincible 'Varun Astra'*. Retrieved May 2025, from Raksha-Anirveda.com: https://raksha-anirveda.com/ unmanned-underwater-vehicle-the-invincible-varun-astra/

Lundquist, C. (. (2023, Q2). *Bluefin Robotics Celebrates its 25th Anniversary*. Retrieved May 2025, from Ocean Robotics Planet: https://www.rovplanet.com/ tportal_upload/ md_publications/rovplanet_35.pdf

Maurya, P. K. (2023, July). C-Bot. *Ocean Digest, 10*.

Sathianarayanan, D. (2013). Deep sea qualification of remotely operable vehicle (ROSUB 6000). *IEEE International Underwater Technology Symposium (UT)*, (pp. 1-7). Tokyo, Japan.

Shome, S., et al. (2010). Autonomous Underwater Vehicle for 150m Depth: Development Phases and Hurdles Faced. In P. e. Vadakkepat, *Trends in Intelligent Robotics. FIRA 2010.* (Vol. 103). Berlin: Springer.

Shome, S., Nandy, S., Das, S., Pal, D., Mahanty, B., Kumar, V., et al. (2010). Autonomous Underwater Vehicle for 150m Depth–Development Phases and Hurdles Faced. In *Communications in Computer and Information Science* (pp. 49-56). Springer Berlin Heidelberg.

Siddiqui, H. (2024, December 2). Indian Navy to Boost Underwater Capabilities with Indigenous AUVs by Sagar Defence Engineering. *Financial Express*.

Singh, R. S. (2023, July 29). India launches 'Neerakshi' - Autonomous Underwater Vehicle for mine detection. *Hindustan Times*.

Skopljak, N. (2022, July 13). *Hugin AUVs bring over •43 million to Kongsberg*. Retrieved May 2025, from Offshore-Energy.biz: https://www.offshore-energy.biz/hugin-auvs-bring-over-43-million-to-kongsberg/

Souza, G. d. (2024, January 9). NIO in Goa launches underwater vehicle C-bot to monitor coral reefs. *Hindustan Times*.

Stommel, H. (1989, April). The Slocum Mission. *Oceanography*.

Taylor, N. E. (2014, April 17). *A robot dives into search for Malaysian Airlines flight*. Retrieved May 2025, from MIT Spectrum.: https://news.mit.edu/2014/robot-dives-search-malaysian-airlines-flight

Tyagi, Y. (2024, November 5). India Making Strides in Autonomous Maritime Security with Matangi USV's 350 Nautical Miles Journey. *Republic World*.

Wadoo, S., & Kachroo, P. (2011). *Autonomous Underwater Vehicles: Modeling, Control, Design, and Simulation*. CRC Press Taylor & Francis Group.

3

Navy Roadmaps and Execution

The Navy is one of the largest stakeholders of underwater technology. Naval applications of unmanned/autonomous technologies include: Underwater survey and repairs; monitoring, surveillance and patrolling, collection of oceanographic data, coordinated search for mines or other submerged objects, and munitions delivery.

The spread of capable autonomous underwater vehicles for scientific and industrial uses has, in turn, encouraged their acceptance and customised their use for naval applications. This chapter traces the evolution of UUVs for naval operations, and how their roadmaps for their development have been drawn up by major navies. Lessons are drawn from the progress made in implementation of such roadmaps by US, China and other nations.

Evolution of UUVs for Naval Applications

Early Developments (1960s–1970s)

Driven mostly by Cold War requirements and the necessity for underwater surveillance and salvage operations, the idea of unmanned undersea vehicles started to take shape throughout the 1960s. Early unmanned systems were often remotely-operated vehicles (ROVs) tethered to surface ships. For example, the US Navy's early experiments with cable controlled underwater recovery vehicles (CURV) in the late 1960s and early 1970s were one of the first systematic attempts to automate underwater operations.[1] These systems

---

[1]   Kaharl, 1990, Streever, 2019

were mostly employed for salvage and recovery tasks as well as for limited mine countermeasure (MCM) operations.

At that time, technological limitations meant that these systems had very limited autonomy. They were controlled directly by operators on board a vessel, with only basic sensor feedback and minimal onboard processing. However, these early platforms laid the groundwork for the independent underwater systems that would follow.

Tethered ROVs to Autonomous Concepts: The 1980s

By the 1980s, improvements in electronics and control systems enabled the development of more capable systems. During this period, navies began deploying systems that—while still largely tethered—incorporated rudimentary forms of autonomy. The US Navy, for instance, used enhanced versions of the CURV systems and began experimenting with autonomous functions for mine countermeasures. Some European navies also started developing their own remotely-operated systems for underwater survey and reconnaissance.

Although these early autonomous functions were limited, they represented a critical transition. The ability to execute pre-programmed missions with minimal intervention was demonstrated in controlled exercises, setting the stage for later developments in truly autonomous underwater vehicles (AUVs).[2]

Emergence of AUVs in the 1990s

The 1990s saw a dramatic transformation in unmanned underwater technology with the emergence of AUVs that could operate independently of a tether. Notable developments during that decade included:

- **REMUS Series:** Developed at Woods Hole Oceanographic Institution (WHOI), the REMUS family of AUVs (including REMUS 100 and REMUS 600) became widely recognized for their capabilities in oceanographic research and mine countermeasures. Although initially designed for scientific applications, the robustness and reliability of REMUS systems soon attracted military interest.

- **Early US Navy AUV Programs:** In the late 1990s, the US Navy began testing AUVs for missions such as mine detection and reconnaissance. These vehicles featured improved onboard processing, sensor

---

[2]   Blidberg, 2001

integration, and autonomous navigation. For example, experimental programs demonstrated the feasibility of AUV deployment from surface vessels and submarines for ASW and MCM tasks.

- **Growing Autonomy and Mission Flexibility:** The idea of executing pre-planned, autonomous missions became more refined. Developers began incorporating path planning algorithms, sensor fusion techniques, and basic decision-making processes that allowed the vehicles to adapt to changing underwater conditions.

These developments marked the start of a trend that would quicken in the next decades: a change from essentially human-operated systems to vehicles capable of autonomous execution of difficult missions.

Integration and Manned–Unmanned Teaming in the 2000s

The new millennium brought significant strides in both hardware and software. In the 2000s, naval forces began to adopt unmanned systems not only for research but also for operational missions.

- **Operational Deployment of AUVs:** Navies of the US, UK, Japan as well as other countries began deploying AUVs for mine countermeasure operations. The REMUS series was further refined and integrated into naval operations. For instance, the US Navy's use of the REMUS 6000—capable of deep-water operations— demonstrated the operational viability of AUVs for ASW and mine detection.

- **Manned–Unmanned Teaming:** A notable development in the 2000s was the integration of unmanned systems with manned platforms. This concept, now widely known as manned–unmanned teaming, involved the coordinated operation of a manned vessel (such as a submarine or frigate) with one or more AUVs. The concept was to extend the situational awareness and operational reach of the manned vessel by using unmanned systems to perform tasks that were too risky, time-consuming, or costly for a human crew.

- **Improved Communication and Data Processing:** With advances in underwater communication and on-board computing, unmanned systems could now share data in real time with their manned counterparts. This enabled more dynamic mission control and improved the overall efficacy of naval operations.

An example from this era is the US Navy's experimentation with AUVs for cooperative mine countermeasures, where a manned vessel would deploy an AUV to scout and neutralize potential mine threats while maintaining overall mission oversight.

The 2010s: Advanced Autonomy, Loitering Munitions, and Networked Operations

The past decade has witnessed further evolution in naval unmanned systems, with breakthroughs in autonomy, artificial intelligence, and networked operations:

- **Advanced Autonomous Capabilities:** In the 2010s, AUVs began to incorporate sophisticated autonomous functions, including adaptive path planning, real-time sensor fusion, and decision-making algorithms. These features enable vehicles to react to dynamic surroundings and unanticipated situations, in addition to executing pre-planned tasks.

- **Loitering Munitions and Strike Capabilities:** A significant new concept has been the development of loitering munitions: unmanned systems that can 'loiter' in an area for prolonged durations and then engage a target depending on conditions. Although loitering munitions have been more commonly associated with aerial systems, naval forces have explored underwater variants capable of covertly monitoring and, if necessary, neutralising enemy assets. Such systems blur traditional lines separating reconnaissance from offensive capability.

- **Manned–Unmanned Teaming 2.0:** The integration between manned and unmanned systems has deepened further. Modern CONOPS involve networks of UUVs operating cooperatively with manned vessels. For example, the US Navy and other allied navies have been developing concepts for large unmanned underwater vehicles (LDUUVs) that can be deployed from submarines or surface ships and work together with manned platforms to enhance anti-submarine warfare (ASW) capabilities.

- **Network-Centric Warfare:** The 2010s also saw the rise of network-centric operations, in which unmanned systems are fully integrated into a digital battle space. Data links, secure communications, and distributed computing allow UUVs to share sensor data, coordinate missions, and contribute to a unified situational picture across a fleet. This trend has been further accelerated by the application of advanced data protocols and 'middleware' frameworks (discussed later).

Examples from this period include the ongoing development and deployment of the REMUS 6000 by the US Navy and its international counterparts, as well as various European and Asian naval programs testing integrated unmanned systems for mine countermeasures, ISR, and even offensive roles.

The 2020s: Current Innovations

The 2020s are witnessing rapid innovations in unmanned naval systems. Key trends include:

- **Swarm Intelligence:** There is a growing interest in using multiple small UUVs operating as a swarm, coordinated by distributed algorithms, to cover large areas, perform surveillance, or engage in complex mine countermeasure operations.

- **Enhanced Manned–Unmanned Teaming:** Recent concepts envision manned platforms (submarines or surface ships) that can control and coordinate a fleet of UUVs in real time, greatly extending their operational reach and flexibility.

- **Loitering and Precision Strikes:** The idea of underwater loitering munitions is gaining traction. Experimental systems have been developed that can remain on station for extended periods, gather intelligence, and then engage targets—either autonomously or under remote human supervision.

- **Artificial Intelligence and Machine Learning:** AI is being deployed to enhance onboard autonomy, enable predictive maintenance, and support rapid decision making. These capabilities allow UUVs to adapt to evolving battlefield conditions and improve mission success rates.

- **Cybersecurity and Resilient Networks:** As unmanned systems become networked, protecting data and control links against cyber threats becomes critical. New standards and protocols are being developed to ensure secure, robust communications among unmanned platforms and with command centres.

Naval programs in several countries including the USA, UK, China and others are actively investing in these technologies. The most significant examples of such deployments have been over the past few years (since 2022) in the Ukraine-Russia war.

US Navy Roadmaps

Several policy documents have been issued over the past two decades by various agencies of the US government and military, outlining visions for induction and utilisation of unmanned systems in war fighting (Figure 3-1).

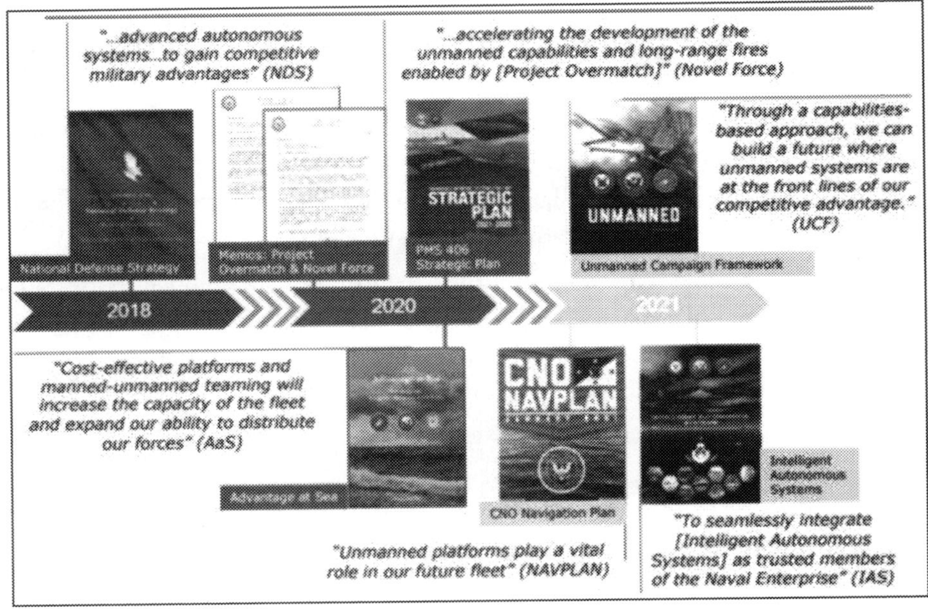

Figure 3-1. Policy documents for Unmanned Systems
(Source: US Navy, 2022[3])

Key Technologies and Responsible Agencies

Advances in Unmanned Underwater Systems (UUS) are driven by several core technologies, each developed or supported by specialised US Navy agencies:

- **Advanced Autonomy & AI:** DARPA (Defense Advanced Research Projects Agency) and ONR (Office of Naval Research) lead research

---

[3]   https://www.navsea.navy.mil/Portals/103/Documents/Exhibits/SNA2022/SNA2022-CAPTPeteSmall-PMS406-LPD-UnmannedMaritimeSys.pdf

into autonomous decision making, sensor fusion, and machine learning for unmanned systems. Their initiatives—such as the unmanned systems integrated roadmap and subsequent programs— aim to reduce operator intervention and enhance swarm capabilities.

- **Propulsion and Energy Systems:** Naval Sea Systems Command (NAVSEA) and the Naval Research Laboratory (NRL) work with industry partners to develop hybrid power systems (e.g., diesel/battery hybrids) and next generation batteries. These efforts target extended endurance, with some systems designed for multi-month missions.

- **Modularity and Payload Integration:** The Naval Sea Systems Command, in coordination with the Naval Supply Systems Command (NAVSUP), is spearheading modular design efforts. These enable rapid reconfiguration of platforms for diverse missions—from ISR to ASW.

- **Underwater Communications and Sensor Networks:** ONR, in collaboration with DARPA, is advancing underwater communication technologies. Initiatives like PLUSNet aim to create resilient acoustic and optical sensor networks that provide robust, real time data connectivity between UUVs and other naval platforms.

- **Cybersecurity and Resilience:** The Defense Information Systems Agency (DISA) and the National Security Agency (NSA) contribute to developing secure communication protocols and protecting unmanned systems against cyber and electronic warfare threats.

These technological drivers form the backbone of the US defence community's long term vision for UUS.

DoD Unmanned Systems Integrated Roadmap

The DoD's integrated roadmap envisions a 25 year evolution of unmanned systems across air, land, and maritime domains. Key objectives include:

- **Interoperability:** Seamless integration among unmanned systems and with manned platforms.

- **Enhanced Autonomy:** Leveraging AI to enable systems to execute complex missions with minimal human oversight.

- **Modularity and Cost Efficiency:** Standardized, reconfigurable platforms that lower acquisition and lifecycle costs.

- **Rapid Fielding:** Accelerated transitions from research and development (R&D) to operational capability, addressing emerging threats in anti-access/area denial (A2/AD) environments.

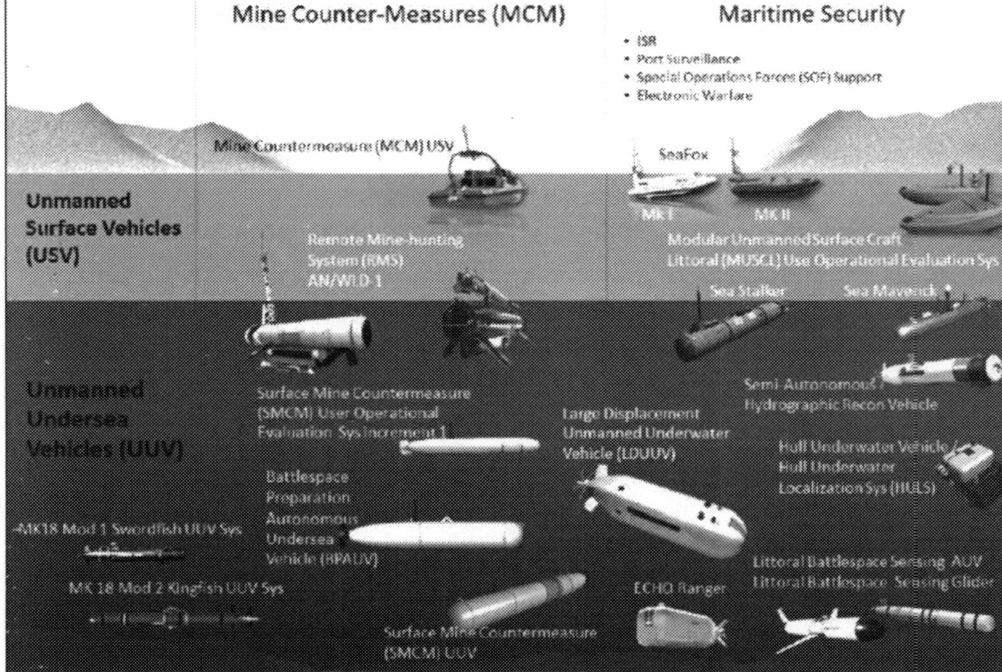

Figure 3-2. Unmanned Maritime Systems categorised by roles (US DoD[4])

For maritime systems, the roadmap specifies roles (e.g., ISR, MCM, ASW) and calls for the development of enabling technologies such as advanced navigation, robust communications, and energy-efficient propulsion systems (Figure 3-2). Agencies like DARPA, ONR, NAVSEA, and NRL play central roles in implementing these objectives.

U.S. Navy's UUV Master Plan

The Navy's UUV Master Plan, first published in 2004 and periodically updated since then, sets forth a vision for integrating unmanned undersea vehicles into a distributed fleet architecture.

Salient elements of the master plan or roadmap (summarised in Figure 3-3) are as follows:

---

4   US Department of Defense, 2013

- **Categorization of UUVs:** Based on displacement and endurance:
  - o **Man-Portable:** (e.g., REMUS 100) for short, tactical missions.
  - o **Lightweight:** (e.g., REMUS 300) for moderate-range ISR and MCM.
  - o **Heavyweight:** (e.g., REMUS 600) for extended ASW and payload delivery.
  - o **Large/Extra-Large:** (e.g., Orca, XLUUV) for long-endurance missions with strategic roles.

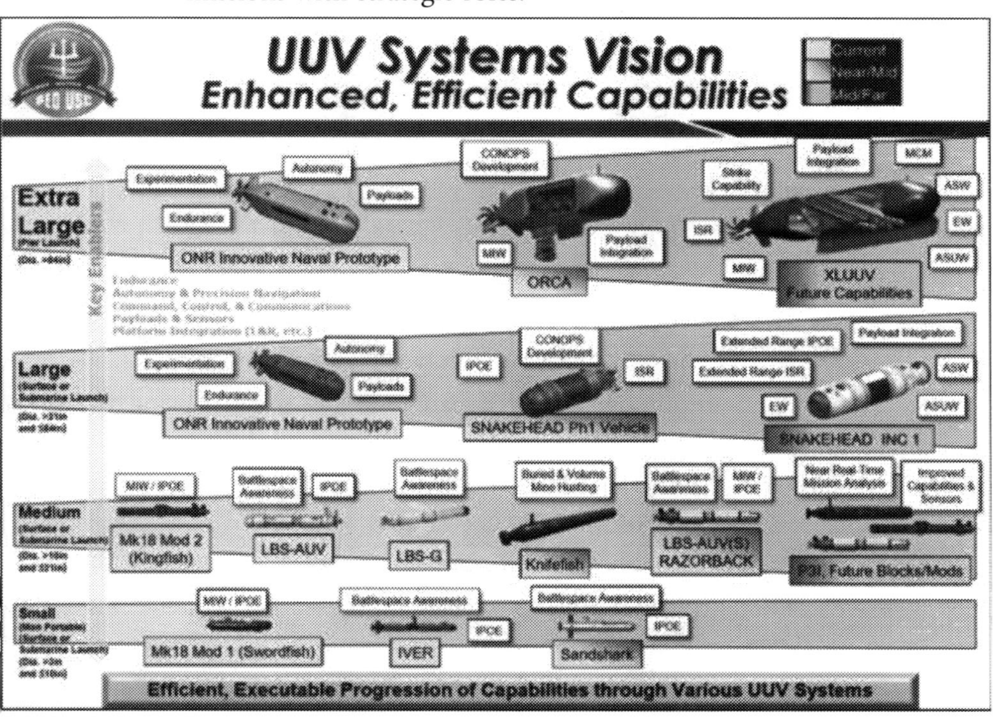

Figure 3-3. US Navy Vision for UUV Systems
(Source: US Navy[5])

- **Mission Roles:** These include ISR, MCM, ASW, covert payload delivery, and underwater logistics.

- **Technology Roadmap:** Detailed development targets for acoustic communications, autonomous navigation, energy storage, and sensor integration.

---

[5] Briefing by Captain Pete Small, Program Manager, Unmanned Maritime Systems (PMS 406) during Surface Navy Association (SNA) 2019 Symposium at Washington D.C. (Vavasseur, 2020).

- **Acquisition Strategies:** Emphasizing rapid prototyping and iterative testing to reduce risk, along with transitioning successful prototypes into production programs.

Agency Roles

NAVSEA oversees shipbuilding and systems integration; ONR supports R&D and sensor network development; DARPA drives breakthrough autonomous capabilities; and the Naval Research Laboratory contributes to advanced sensor and communication technologies. The program executive office (PEO) for unmanned and small combatants (USC) of the NAVSEA is responsible to design, develop, build, maintain and modernize the US Navy's unmanned maritime systems; mine warfare systems; special warfare systems; expeditionary warfare systems; and small surface combatants.

Salient Trends and Updated Targets

Common trends across these roadmaps include:

- **Modularity and Scalability:** Platforms must be reconfigurable, allowing rapid changes in mission payloads.

- **Increased Autonomy:** There is a marked shift towards reducing human oversight through AI, with updated targets for fully autonomous operation in contested environments.

- **Cost-Effective Production:** Accelerated acquisition strategies now emphasize 'attritable' systems produced at lower unit costs.

- **Extended Endurance:** New targets call for continuous operation over several months and operational ranges spanning thousands of nautical miles.

- **Joint Integration:** Unmanned systems are being developed as integral nodes in network-centric warfare, requiring robust sensor networks and data integration protocols.

The acquisition approach has shifted over time—from lengthy, risk-averse programmes to agile, iterative processes that emphasize rapid prototyping, production readiness reviews, and adaptive acquisition frameworks (e.g., Section 804 authority and the Replicator initiative).[6] These changes are

---

[6]   Shah & Kirchhoff, 2024

designed to address the urgent need for new capabilities in a rapidly evolving threat environment.

Programme Targets and Progress

Key US Navy programme timelines and their progress are detailed below.

- **Orca (XLUUV):**
  - o *Timeline:* Initially targeted for delivery by end-2022; revised targets now project full delivery between February and June 2024.
  - o *Progress:* The first Orca was delivered in December 2023. Its modular design and hybrid power system have extended endurance for several months.
- **Snakehead (Large Displacement UUV):**
  - o *Timeline:* Under development with strategic milestones set for prototype testing and evaluation within the next two years.
  - o *Progress:* Early prototypes are in evaluation; detailed timelines continue to be refined.
- **LDUUV (Large Displacement UUV):**
  - o *Timeline:* Planned procurement beginning in the mid-2020s for roles in strategic mine deployment and ASW.
  - o *Progress:* Several prototypes have been tested; acquisition decisions are pending further risk and performance assessments.
- **REMUS Family (Small to Medium UUVs):**
  - o *Timeline:* Multiple models (REMUS 100, 300, 600) are operational, with ongoing upgrades to extend endurance and sensor capabilities.
  - o *Progress:* Widely deployed across various missions; continuous field improvements are reported.

Table 3-1 summarises UUV categories as per the US Navy Roadmap and their roles, planned timelines, and progress.

Table 3-1. US Navy UUV Roadmap Targets and Progress

Category	Example Program(s)	Primary Roles/ Missions	Planned Timeline/ Target	Current Progress
Small	REMUS 100, M3V	ISR, mine counter-measures, rapid environmental assessment	Fielded; operational since 2010–2015	Widely deployed; iterative upgrades ongoing
Medium	REMUS 300, Kingfish (Mk 18 Mod 2)	ISR, mine detection, moderate-range missions	Prototypes validated; transition as SUUV aimed by 2023	Production of Kingfish completed. Trials for successor Viperfish, ongoing
Large	Snakehead LDUUV	Extended ASW, covert payload delivery, deep-sea operations	Snakehead induction aimed 2025	Significant technical/cost issues; program cancellation planned (2022) but limited prototype testing resumed in 2024.
Extra Large	Orca XLUUV	Strategic mine deployment, long-endurance, covert missions	Delivery initially targeted for 2020	First delivery Dec 23; >3 years delay, 64% ($242M) over budget

Sensor Networks and Naval Exercises

Efforts to develop robust underwater sensor networks have been driven by agencies such as DARPA and ONR. Programs like PLUSNet aimed to create resilient, multi modal sensor networks that provide continuous, high-fidelity communication among UUVs, manned platforms, and command centres. These networks are critical for:

- **Real-Time Data Fusion:** Enabling rapid decision-making in complex operational environments.

- **Swarm Coordination:** Facilitating the synchronized operation of unmanned vehicles in distributed fleet architectures.

Naval exercises have already incorporated these developments:

- **Digital Talon Exercises:** The Navy and Marine Corps have used prototype unmanned systems in exercises to test autonomous navigation and sensor networking.

- **Project Convergence:** Joint exercises that integrate unmanned surface and undersea vehicles with manned platforms to validate network-centric warfare concepts.

- **ACTUV Sea Trials:** Demonstrations of DARPA-funded unmanned vessels like Sea Hunter have provided valuable operational data that feeds back into refining UUV capabilities and acquisition strategies.

Implementation of Roadmaps

Congressional Report Insights

Congressional Research Service (CRS) reports emphasize that the Navy's ambitious goal of integrating large numbers of unmanned systems into a distributed fleet architecture remains transformative.[7] They note that:

- **Ambitious Goals:** Fleet architecture plans include hundreds of unmanned systems to supplement a reduced number of manned platforms.

- **Accelerated Acquisition Risks:** While rapid acquisition aims to field capabilities quickly, it also brings risks in terms of insufficient risk management and integration challenges.

- **Industrial Base Concerns:** There is growing concern about whether the defence industrial base can scale production to meet the ambitious targets.

GAO Findings and Recommendations

US Government Accountability Office (GAO) assessments through two reports of 2022[8] identified the following challenges in US Navy unmanned programmes:

- **Cost Overruns and Delays:** Programs like the XLUUV have experienced significant cost overruns (up to 64%) and schedule delays (over three years), attributed to incomplete production readiness reviews and insufficient cost/schedule analyses.

- **Fragmented Management:** The absence of an integrated portfolio management approach has hampered coordinated progress across multiple programs.

---

[7] O'Rourke, March 2025

[8] US Government Accountability Office, April 2022; US Government Accountability Office, September 2022.

- **Insufficient Risk Assessments:** GAO recommends that the Navy institute robust risk assessments and detailed business cases to improve decision-making and mitigate further delays.

- **Transition Challenges:** The Navy's rapid prototyping strategy needs a clearer path for transitioning prototypes to full-scale production.

Chinese Developments

China is aggressively advancing its unmanned and autonomous underwater capabilities, driven by the aim to counterbalance US naval superiority and to secure its strategic maritime interests. Facing challenges in resource exploration and maritime security, the People's Republic of China (PRC) has invested heavily in underwater technologies. These efforts span a wide range of platforms including AUVs, ROVs, manned submersibles, and sophisticated underwater sensor networks.

China's integrated approach—involving both state-owned enterprises and academic–military R&D collaborations—is reshaping underwater capabilities and influencing the regional security balance. Several key agencies and institutions drive underwater systems development in China, as follows:

- **China Shipbuilding Industry Corporation (CSIC) and China State Shipbuilding Corporation (CSSC):** Two of the largest state-owned enterprises responsible for the design, construction, and modernization of naval vessels and underwater platforms. They are heavily involved in the development of both manned and unmanned underwater systems.

- **Chinese Academy of Sciences (CAS):** CAS, through its various institutes (e.g., the Institute of Underwater Engineering at Shanghai Jiao Tong University, and the Shenyang Institute of Automation), plays a pivotal role in fundamental research, prototype development, and system integration for AUVs, ROVs, and sensor networks.

- **Ministry of National Defence (MOD) and People's Liberation Army (PLA) Naval Research Institutions:** These agencies drive military-specific research. The PLA Navy (PLAN) emphasizes the development of AUVs for mine countermeasures and anti-submarine operations, while also investing in underwater sensor networks to support surveillance and intelligence.

- **China Electronics Technology Group Corporation (CETC):** CETC contributes to the development of advanced sensor and communication technologies that are crucial for underwater systems.
- **State Key Laboratories:** Several state key laboratories, such as those at Harbin Engineering University and Shanghai Jiao Tong University, focus on underwater robotics, energy systems, and control algorithms.

Each of these organizations contributes to a multifaceted development effort that combines advanced R&D with production capabilities, enabling China to field a broad range of underwater platforms.

Chinese Underwater Systems Roadmaps

Chinese government documents and academic publications have laid out long-term roadmaps for underwater systems development. The latest Annual US DoD Report on Military Capabilities of China states: "By 2030, the PLA expects to field a range of 'algorithmic warfare' and 'network-centric warfare' capabilities operating at different levels of human-machine integration. CCP leaders believe AI and machine learning will enhance information, surveillance, and reconnaissance capabilities and enable a range of new defence applications, including autonomous and precision-strike weapons. The PRC is invested in autonomous vehicles, predictive maintenance and logistics, and automated target recognition. The PLAN is interested in unmanned underwater vehicles, similar to the PRC's commercial unmanned boats".[9]

In an interview in July 2018, Lin Yang, the director of the Shenyang Institute of Automation noted that China plans to develop new-generation military underwater robots by 2021 to coincide with the 100-year anniversary of the Chinese Communist Party in 2021. The goal of the 912 Project is to develop new-generation military underwater robots, particularly AI-driven unmanned submarines, to handle surveillance, mine laying, and attack missions. Likewise, Luo Yuesheng, professor at the College of Automation in Harbin Engineering University, a major development centre for China's new submarines, contends that "AI subs would put the human captains of other vessels under enormous pressure in battle…It is not just that the AI subs are fearless, but that they could learn from the sinking of other AI vessels and adjust their strategy continuously. An unmanned submarine trained to be familiar to specific water will be a formidable opponent".[10]

---

9   US Department of Defense, 2024

10   Sakhuja, 2019

Other Chinese references—found in journals such as *Robotics Technology and Applications* and *Underwater Robotics* emphasize the importance of modularity, high endurance, and the integration of advanced sensors and AI in underwater platforms. These publications serve as the Chinese roadmap for both civilian and military applications and stress the need for robust underwater sensor networks and cooperative multi-vehicle operations.[11]

Key thrust areas in China's underwater technology include:

- **Stealth and Endurance:** Development of advanced UUVs with extended operational ranges, improved stealth features, and high-resolution sensor arrays.

- **Underwater Sensor Networks:** Significant investments in integrated underwater sensor systems, enabling real-time data collection and swarm coordination. Chinese research emphasizes the fusion of AI with advanced acoustic and optical communication to create resilient networks.

- **Weaponisation and ASW:** Incorporation of advanced missile systems and electronic warfare capabilities into UUV platforms to enhance anti-submarine and anti-surface warfare.

- **Cost-Effective Production:** Focus on rapidly scaling production using domestic suppliers and leveraging commercial off-the-shelf (COTS) technologies to keep costs low while maintaining technical superiority.

Major Products and Capabilities

China has demonstrated a broad range of underwater platforms, many of which are at the forefront of global technological development. Below, we detail key developments across AUVs, ROVs, and manned submersibles, as well as underwater sensor networks.

Autonomous Underwater Vehicles (AUVs)

Chinese AUVs have been developed under various national R&D programs (such as the 863 and 912 programmes) and include several notable classes:

- **Explorer AUV:** The first Chinese AUV developed under the 863 Program, designed for deep-sea mapping and environmental monitoring. It laid the foundation for subsequent designs.

---

[11] Fedasiuk, 2021

Figure 3-4. Various Chinese AUVs and ROVs
(Source: Jianguo Wu[12])

- **CR Series (e.g., CR 01, CR 02):** These vehicles have been developed with improved positioning, modular payloads, and enhanced endurance. They are used for both scientific research and mine countermeasure missions.

- **Sea Whale Series:** Designed for long-range, deep-sea exploration, these AUVs feature extended endurance and high-resolution mapping capabilities.

- **Micro Dragon Series:** These are smaller, more agile AUVs developed for confined-space operations and rapid deployments.

- **Wukong AUV:** A next-generation AUV designed for ultra-deep operations, reportedly capable of diving beyond 10,000 metres.

---

[12] Jianguo, 2018

Table 3-2. Characteristics of various Chinese AUVs

AUV Model	Length (m)	Max Depth (m)	Endurance	Primary Roles
Explorer	~4.4	6000	10–15 hours	Seafloor mapping, environmental monitoring
CR-01	~4.4	6000	10–20 hours	Mine countermeasures, ISR, coastal survey
CR-02	~4.5	6000	25 + hours	Deep-sea surveying, mineral exploration
Sea Whale 1000	~?	1500–2000	>1 month	Long-range mapping, resource exploration
Sea Whale 2000	~?	1500–2000	>1 month	Extended deep-sea exploration, data collection
Micro Dragon Series	~1.5–2.5	1000–3000?	Several hours	Confined space operations, rapid response
Wukong	~2.0	>10,000	Extended endurance	Ultra-deep exploration, strategic ISR

Remotely Operated Vehicles (ROVs)

Chinese ROV development has focused on systems that support both scientific research and military operations:

- **Haiyan ROV:** A mid sized ROV used for underwater inspection, salvage, and scientific exploration.

- **Deep Discoverer Series:** ROVs that combine high-resolution imaging with advanced sensor capabilities for detailed seafloor analysis.

- **Specialised Military ROVs:** Adapted for mine countermeasure tasks and surveillance, often integrated with autonomous systems for limited human intervention.

Manned Submersibles

China is renowned for its manned deep-submergence vehicles, which have set several records (Figure 3-5, Table 5-3):

- **Fendouzhe (Striver):** One of the deepest-diving manned submersibles, capable of reaching depths of over 10,000 metres. It is primarily used for scientific research and resource exploration.

- **Jiaolong:** A manned submersible developed for both research and technology demonstration, widely used in deep-sea missions.

- **Shenhai Yongshi:** Another advanced manned submersible noted for its operational reliability and versatility in diverse underwater missions.

Figure 3-5. Chinese Submersibles and Large AUVs
(Source: Jianguo Wu[13])

Table 3-3. Summary of Chinese Manned Submersibles

Submersible Model	Max Depth (m)	Crew Capacity	Primary Roles
Fendouzhe (Striver)	>10,000	3–4	Deep-sea exploration, scientific research, mineral exploration
Jiaolong	~7,000–8,000	3	Deep-sea research, marine geology, biological studies
Shenhai Yongshi	~6,000–7,000	2–3	Inspection, research, underwater archaeology

Underwater Sensor Networks

China is also investing in underwater sensor networks to support both civilian and military applications.[14] These networks aim to provide continuous, high-bandwidth communications between underwater vehicles and command centres. Key initiatives include:

- **Integrated Underwater Sensor Arrays:** Developed by CAS and affiliated research institutes to enable real-time data collection for

---

[13] Jianguo, 2018
[14] Dahm, 2020

underwater surveillance, environmental monitoring, and mine countermeasure operations.

- **PLUSNet-Like Systems:** Efforts similar to the US DARPA/ONR PLUSNet project are underway in China under various research initiatives within the Ministry of Science and Technology and CAS to integrate acoustic, optical, and electromagnetic sensors into cohesive networks.

- **Joint R&D Programs:** Collaborative projects between state-owned enterprises (CSIC, CSSC) and academic institutions have accelerated the development of advanced underwater communications, navigation, and sensor fusion technologies.[15]

In March 2025, it has been reported that China had begun construction of a deep-sea research platform in Guangzhou to explore extreme marine environments. The Research Facility of the 'Cold-seep Ecosystem' is led by the South China Sea Institute of Oceanology and is scheduled for completion within five years. The project will combine a manned deep-sea laboratory on the ocean floor with advanced land-based simulation systems. The design is intended to allow long-term, high-precision studies of cold-seep ecosystems— unique biological communities that thrive in darkness and under extreme pressure, where methane and other chemicals seep from the seafloor.[16]

Deployments and Military Exercises

Chinese underwater systems have been deployed in various operational and experimental contexts:

- **Scientific Missions:** Manned submersibles like Fendouzhe have been used in deep-sea exploration missions in the Mariana Trench, while AUVs such as the Explorer and CR series have conducted extensive mapping and environmental surveys in the South China Sea.

- **Military Exercises:** The People's Liberation Army Navy (PLAN) has conducted numerous exercises in the South China Sea that incorporate unmanned systems. For instance, exercises simulating mine countermeasure operations have deployed AUVs alongside manned platforms to test interoperability and real-time sensor networking.

---

[15] Agnihotri, 2019
[16] Dongjie, 2025

- **Joint Demonstrations:** During multi-domain naval exercises, Chinese forces have integrated unmanned systems with satellite and aerial surveillance to enhance underwater situational awareness. These exercises have showcased capabilities such as coordinated autonomous swarming and rapid target engagement.

Chinese strategic documents and military exercises suggest that China is not only fielding advanced unmanned underwater platforms but also integrating these systems into larger naval and joint force operations. Their roadmap indicates a long-term push towards achieving a distributed underwater force capable of operating in contested A2/AD environments.

Other Nations

India's Approach

Initial development of AUVs in India was discussed in Chapter 2. In 2021, the Indian Navy brought out a comprehensive wish-list called 'Integrated Unmanned Roadmap'. This aims to provide a comprehensive Unmanned Systems Roadmap in consonance with the Indian Navy's Concept of Operations and chart out a capability development plan.[17] The document describes the applications and solutions envisaged for autonomous/remotely-operated vehicles for air, surface and sub-sea use. The document is available to Indian industry on request from the Directorate of Staff Requirements at Naval Headquarters, and is intended to promote the vision of *Atmanirbhar Bharat*.

Other documents and mechanisms that share the government and Navy's future requirements are the Indian Naval Indigenisation Plan and the Positive Indigenisation Lists issued by the MoD. Outreach events are organised regularly by the Navy, often through the Society of Indian Defence Manufacturers (SIDM).

India's unmanned systems development, aligned with its strategic imperatives, is likely to include the following key aspects:

- **Maritime Domain Awareness:** Prioritising unmanned systems for ISR and MCM.

---

[17] https://www.pib.gov.in/PressReleasePage.aspx?PRID=1764695; Joseph, 2023.

Figure 3-6. Indian Navy's Integrated Unmanned Roadmap released in 2021[18]

- **Indigenous Capability:** Emphasis on indigenous technology by fostering domestic R&D and public–private partnerships.
- **Timelines:** Near-term operational capabilities are targeted, with vision of integrating UUVs with manned naval platforms.[19]

Defence Innovation Ecosystem

The recent impetus given to the defence innovation ecosystem is a success story worth celebrating. 'Innovations for Defence Excellence' (iDEX) is an initiative and operational framework launched in April 2018 to foster innovation in the defence sector by connecting innovators with the armed forces. It creates an ecosystem for startups, MSMEs, and individual innovators to develop technology for national defence and security.[20]

The iDEX framework is funded and managed by the Defence Innovation Organisation (DIO) under the aegis of the Department of Defence Production (DDP), Ministry of Defence, Government of India. The DIO is a Section-8 (not-for-profit) company, formed, funded and managed by Hindustan Aeronautics Limited (HAL) and Bharat Electronics Limited (BEL).[21]

The iDEX provides grants to Start-ups/MSMEs for projects in many technological areas, under Defence India Start-up Challenges (DISC) and Open Challenge, through the Support for Prototype and Research Kickstart

18 https://psuwatch.com/topic/integrated-unmanned-roadmap

19 Agnihotri, 2019

20 https://www.pib.gov.in/Pressreleaseshare.aspx?PRID=1541026#:

21 https://www.myscheme.gov.in/schemes/idex

(SPARK) Framework.[22] The grants for each project can be up to Rs. 1.5 CR, Rs. 10 CR (iDEX Prime) and even Rs. 25 CR (iDEX ADITI)

The iDEX approach, besides fostering innovation and technology development, is also a path to procurement for the Armed Forces as per the Defence Acquisition Procedure 2020, which enables procurement through innovative solutions under the iDEX or the Technology Development Fund (TDF) Scheme of the DRDO.

Figure 3-7. Event held in October 2024 to felicitate iDEX challenge winners[23]

The Indian Navy's SPRINT initiative was launched through iDEX in July 2022.[24] This was driven by the Navy's Technology Development and Acceleration Cell (T-DAC). The project name 'SPRINT' was coined to denote 'Supporting Pole-Vaulting in R&D through iDEX, NIIO and Technology Development Acceleration Cell (TDAC)'. Under the SPRINT project, for 75 technology challenges proposed by the Navy, over 1,100 proposals were received from MSMEs across the country. About 115 SPRINT contracts were signed with Indian industries in the first year itself (2022-23).[25] These SPRINT

---

22   https://www.ddpmod.gov.in/offerings/schemes-and-services/idex

23   https://x.com/India_iDEX/status/1843910125511422140/photo/2

24   https://www.pib.gov.in/PressReleasePage.aspx?PRID=1842449

25   https://www.rajnathsingh.in/press-release/indias-defence-sector-is-riding-on-the-boat-of-innovation-shri-rajnath-singh/

projects have covered various emerging areas such as AI-powered maintenance, underwater drones, and intelligent mobility solutions. With the close involvement of the end-user, most of the projects transitioned to successful prototypes, with several graduating to production orders (Figure 3-7).[26]

There have been significant successes,[27] notwithstanding concerns about the long-term viability and in-service supportability of these products. The innovations and accelerated developments have been enabled by driven individuals among end-users leveraging these newly-created mechanisms.

Collaborative Efforts for Design

The Indian Navy's Submarine Design organisation (DND(SDG)) has been collaborating with several academic institutions, industry partners and R&D labs of the GoI to advance underwater technologies and underwater vehicles. The organisation has developed in-house designs of AUVs, filed patents, and steered R&D studies for design, analysis and development of AUV systems and technologies through various national institutions. Enabling frameworks (MoUs) have been formalised with several institutions and agencies.

There have been fruitful and continuing interactions with experts in NIO Goa, NIOT Chennai, TIH-IoT (IIT Bombay), IIT Madras and IIT Gandhinagar. Dual-use projects for technology development and demonstration in the field of Marine IoT are being funded by TIH-IoT of IIT Bombay. A test bed AUV *Jaya* has been developed by NIO Goa for refining sensors and software. (Figure 3-8)

All these inputs have been used by DND(SDG) for its in-house design of an AUV of the extra-large category, whose design has been named *Jalkapi* (Sanskrit for 'dolphin').

---

[26]   https://www.financialexpress.com/opinion/traversing-the-innovation-frontier-the-impact-of-idex-sprint-on-our-growth-trajectory/3417300/ (Verma, 2024)

[27]   https://www.ddpmod.gov.in/offerings/schemes-and-services/idex/products

Figure 3-8. AUV 'Jaya' developed by NIO Goa

Figure 3-9. Navy's design of AUV 'Jalkapi' showcased in October 2024[28]

A contract for building a prototype of the Navy-designed AUV 'Jalkapi' (Figure 3-9) was awarded under the iDEX-ADITI scheme to M/s RekiSe Marine in 2024.[29] Structural construction of this prototype, which would be the largest AUV to be built in India, was started through M/s Krishna Defence and Allied Industries Ltd (KDAIL) in Gujarat in June 2025 (Figure 3-10).[30]

---

28 https://www.instagram.com/p/DBs9XzAtc89/?locale=zh-hans&img_index=3

29 https://x.com/Varun55484761/status/1819392445152743861

30 Construction of India's Largest Unmanned Submarine 'Jalkapi' Begins in Halol, Gujarat, 2025

Figure 3-10. Steel cutting function for AUV 'Jalkapi' in June 2025[31]

European and Australian Initiatives

European navies (e.g., that of the UK, France, Germany) and Australia are also advancing their unmanned systems. These initiatives stress interoperability and regional defence collaboration, with NATO allies investing in standardized unmanned systems to bolster collective maritime security.

- **Europe:** Focus on interoperable systems that work within NATO frameworks and address regional maritime threats.

- **Australia:** Significant investments in unmanned systems under initiatives like AUKUS, aiming to rapidly field autonomous vessels to counter regional challenges.

Future Directions and Recommendations

Areas for Improvement

To fulfil the ambitious goals of naval roadmaps, enhancements are needed:

- **Unified portfolio management:** Develop an integrated management framework that consolidates unmanned systems programs across

---

31   https://deshgujarat.com/2025/06/11/construction-of-indias-largest-unmanned-submarine-jalkapi-begins-in-halol-gujarat/

various government agencies and stakeholders for coordinated resource allocation.

- **Make trial facilities available:** Shared use of trial facilities (lake, harbour, sea) for all developers to rigorously test their prototypes, without small firms having to invest in developing such infrastructure.

- **Standardization and interoperability:** Development of common technical standards and communication protocols for seamless integration with manned platforms.

- **Speedy acquisition practices:** Acquisition approvals, orders and payments needs to be speedily executed to facilitate small technology firms to remain solvent and to encourage rapid prototyping and iterative testing.[32]

Strategic and Operational Recommendations

- **Integrate development ecosystem:** Establish a coordinated technology strategy that bridges unmanned systems development efforts across various national agencies and unmanned applications.

- **Invest in advanced autonomy:** Prioritize R&D in AI and machine learning to enable higher levels of autonomous operation, particularly in contested environments.

- **Expand industrial base:** Broaden the defence industrial base by including non-traditional partners and incentivising rapid production.

- **Simplify acquisition processes:** Transform rules as well as mindsets in financial processes. Tech firms demonstrating successful prototypes (through mechanisms like DIU in US or iDEX in India) need to be guaranteed production orders, without requiring them to compete in another bidding process.[33] Procurement or acquisition processes need to be simplified, with adaptive funding models to encourage innovation and cut down timelines.[34]

- **Multi-agent integration:** Enhance network-centric capabilities to ensure that unmanned systems become fully integrated nodes in joint and multi-domain operations.

---

[32] Kumar, 2025

[33] Shah & Kirchhoff, 2024

[34] Kumar, 2025

- **Underwater sensor networks:** Scale up projects to improve underwater communications, sensor fusion, and swarm coordination.
- **Validate through naval exercises:** Continue and expand trials facilities and organisations within the Navy. Conduct naval exercises repeatedly to validate new capabilities of uncrewed systems in realistic operational environments.

Conclusion

The next decade promises significant advances in unmanned underwater technology driven by improvements in autonomy, energy systems, sensor integration, and modular design. However, realising these advancements requires overcoming current challenges in cost management, integrated oversight, and risk assessment.

Technology and capability roadmaps of the US set forth ambitious visions for a technologically advanced maritime force with a diverse range of platforms and capabilities, emphasising modularity, enhanced autonomy, interoperability, and rapid fielding of capabilities.

China is also aggressively pursuing underwater technology, focusing on stealth, endurance, and networked sensor capabilities. India has been progressing with emphasis on indigenous development but cautious inductions.

To achieve the strategic objectives outlined in such roadmaps, it is necessary to revamp acquisition strategies, encourage rapid prototyping, place bulk production orders, and foster joint civil-military collaboration. The acquisition approach needs to shift from long-term, rigid programs to agile, iterative processes that emphasise rapid prototyping and encourage successful development/ demonstration by quickly placing production orders.

With these adjustments, the vision of a fleet augmented by hundreds of advanced unmanned underwater systems can be realised, ensuring maritime superiority in an increasingly contested and dynamic global environment.

While formulation of a roadmap requires vision, its realisation requires technological capability. In the next chapter we dive into the major components and design processes of AUVs, and their hardware and software architecture.

References

Agnihotri, K.K. (2019). High Technology Developments in China: Leveraging for Military Effectiveness. Centre for Joint Warfare Studies. Retrieved from CENJOWS: https://cenjows.in/wp-content/uploads/2022/03/High-Technology-Developments-by-Capt-IN-KK-Agnihotri.pdf

Blidberg, D.R. (2001). The Development of Autonomous Underwater Vehicles (AUV): A Brief Summary. IEEE International Conference on Robotics and Automation. Seoul, Korea.

Dahm, J.M. (2020, June 16). Exploring China's Unmanned Ocean Network. Asia Maritime Transparency Initiative. Retrieved from AMTI: https://amti.csis.org/exploring-chinas-unmanned-ocean-network/

Dongjie, Y. (2025, March 2). *China begins construction on deep-sea research facility to study cold-seep ecosystems.* Retrieved from ChinaDaily: https://www.chinadaily.com.cn/a/202503/02/WS67c4546aa310c240449d8116.html

Fedasiuk, R. (2021, August 17). Leviathan Wakes: China's Growing Fleet of Autonomous Undersea Vehicles. Centre for International Maritime Security. Retrieved from CIMSEC: https://cimsec.org/leviathan-wakes-chinas-growing-fleet-of-autonomous-undersea-vehicles/

Jianguo, W. (2018). Analysis of the current status and problems of the development of China's autonomous underwater robots and core components. *Oceanology International.* Hebei University of Technology. Retrieved from https://www.oceanologyinternational.com/content/dam/sitebuilder/rxch/oichina/document/Autonomous%20and%20Remotely%20Operated%20Underwater%20Vehicles%20and%20Vessels_Development%20and%20problem%20Analysis%20on%20autonomous%20underwater%20veh

Joseph, B.S. (2023, August 7). Taking Stock of India's Evolving Unmanned Undersea Capabilities. *The Diplomat.* Retrieved from The Diplomat: https://thediplomat.com/2023/08/taking-stock-of-indias-evolving-unmanned-undersea-capabilities/

Kaharl, V. A. (1990). *Water Baby: The Story of Alvin.* New York: Oxford University Press.

Kumar, Ajay (2025, September 11). Rewiring public procurement for innovation faces single-vendor hurdle. *Business Standard.* Retrieved from: https://www.business-standard.com/opinion/columns/rewiring-public-procurement-for-innovation-faces-single-vendor-hurdle-125091101408_1.html

O'Rourke, R. (March 2025). *Navy Large Unmanned Surface and Undersea Vehicles: Background and Issues for Congress.* Washington DC: Congressional Research Service. Retrieved from https://www.congress.gov/crs-product/R45757

Sakhuja, V. (2019, December 9). *China's Underwater Initiatives and Seabed Exploration Aspirations.* Retrieved from Def Strat: https://www.defstrat.com/magazine_articles/chinas-underwater-initiatives-and-seabed-exploration-aspirations/

Shah, R. M., & Kirchhoff, C. (2024). Unit X: How the Pentagon and Silicon Valley Are Transforming the Future of War. Scribner.

Shuo, L. & Hongyu, Z. (2022). Application and prospect of unmanned underwater vehicle. *Bulletin of Chinese Academy of Sciences, 37*(7), 910-920. Retrieved June 2025 from https://bulletinofcas.researchcommons.org/cgi/viewcontent.cgi?article=2030&context=journal

Streever, B. (2019). *In Oceans Deep: Courage, Inovation and Adventure Beneath the Waves.* New York: Little, Brown and Company.

US Department of Defense. (2011). *Unmanned Systems Integrated Roadmap FY2011-2036.* Office of the Under Secretary of Defense (Acquisition Technology and Logistics). Retrieved from https://apps.dtic.mil/sti/pdfs/ADA558615.pdf

US Department of Defense. (2013). *Unmanned Systems Integrated Roadmap FY2013-2038.* Washington DC: Office of the Under Secretary of Defense (Acquisition Technology and Logistics). Retrieved from https://ntrl.ntis.gov/NTRL/dashboard/searchResults/titleDetail/ADA592015.xhtml

US Department of Defense. (2018). *Unmanned Systems Integrated Roadmap FY2017-2042.* Washington DC: Office of the Under Secretary of Defense (Acquisition Technology and Logistics). Retrieved from https://apps.dtic.mil/sti/citations/AD1059546

US Department of Defense. (2024). *Military and Security Developments Involving the People's Republic of China.* Annual Report to Congress, Washington DC.

US Government Accountability Office. (April 2022). *Uncrewed Maritime Systems: Navy Should Improve Its Approach to Maximize Early Investments.* Retrieved from https://www.gao.gov/products/gao-22-104567

US Government Accountability Office. (September 2022). *Extra Large Unmanned Undersea Vehicle: Navy Needs to Employ Better Management Practices to Ensure Swift Delivery to the Fleet.* Retrieved from https://www.gao.gov/products/gao-22-105974

Vavasseur, X. (2020, April 2). *US Navy Issues Draft RFP for the MUUV Program.* Retrieved from Naval News: https://www.navalnews.com/naval-news/2020/04/u-s-navy-issues-draft-rfp-for-the-muuv-program/

4

AUV Design Approach

The design of a product is the ultimate application of an engineer's expertise. AUV design is a multi-disciplinary endeavour that spans mechanical engineering, electronics, software, and control systems. The relatively small physical scale of an Autonomous Underwater Vehicle (AUV) compared to a submarine makes it a manageable project for a small team of engineers. And if the team is efficient (and fortunate!), within a few months they can make their creation float, dive and surface.

AUVs must operate in complex, dynamic underwater environments, which demand that all components are tightly integrated: from the hull and propulsion system to the sensor suite and onboard computers. The design must satisfy several conflicting requirements: high energy efficiency, robust operation under high pressure, low maintenance, and high autonomy. This chapter introduces the design process for an AUV in qualitative terms, to the multi-faceted engineer who aspires to be a designer.

This chapter starts with the possible approaches for initial size estimation for a new design. We then discuss the major components and systems of an AUV including sensors, payload, structures, hydrodynamics and control, vehicle systems, and software functions. We briefly look at Systems Engineering as an approach for AUV design, and then outline salient aspects of AUV hardware and software layers.

Vehicle Sizing Approaches

The traditional approach for ship or submarine design starts with the formulation of requirements, followed by progressive estimation of main parameters and performance, with iterations undertaken to analyse and describe each of the sub-systems in further detail.

Context

Development of submersibles in the latter half of the 20th century resulted in a variety in external form and capabilities, but with the underlying necessity of a pressure-proof capsule or sphere to enclose a human crew. Un-crewed solutions started with assemblies towed underwater for scanning or taking photographs or undertaking manipulation (lifting) of objects on the sea bed. Un-tethered and un-crewed craft were developed initially as kinetic weapons (such as a torpedo). However, UUVs as we recognise them today, entered service as slow-moving, man-portable vehicles for benign scientific applications. Another line of development evolved from buoys and vertical samplers by adding motion or propulsion to sea water sampling devices.

For all types of ships (including unconventional semi-displacement craft, warships and submarines), the responsibility of design has been that of a professional known as Naval Architect. Dependence on past experience would vary as per the extent of novelty of the vessel. Iterations become essential in the design process because changes in one aspect of the design affect several others. What has remained common is the overall approach of progressive detailing of the design solution with each iteration.

This role of synthesis and the approach adopted for ship or submarine design can been generalised for any complex technological product development through the parlance of 'Systems Engineering'.

We start by discussing alternative approaches to estimate the initial size of a new UUV to be designed.

Parent-Ship Approach

AUVs and ROVs in service at present have a variety of sizes and capabilities, affording today's designer the possibility of the 'parent ship' approach that naval architects traditionally adopt for ship design.

The parent-ship approach essentially involves scaling of an existing design to meet modified requirements of a new design. The method works best when changes are slight, say of the order of 10 per cent for a major feature. The approach becomes tricky when a parameter such as speed may need to be increased by (say) 10 per cent, while another parameter such as endurance may need scaling by a different factor (say 50%). Each parameter may have different correlations for scaling, which need to be physically consistent. For example, power requirement is proportional to the cube of speed. Internal volume, affecting fuel/energy content, is proportional to the cube of linear dimension. However, change in form would affect drag, power consumption and endurance in ways that need to be considered in terms of physical effects on skin friction drag as well as viscous pressure drag, and may not be amenable to scaling by simple ratios.

Parametric Studies for Design Exploration

The advantage of the essentially axisymmetric form of AUVs is that it is possible to geometrically estimate the form volume, and thus estimate the resistance and power requirements using empirical methods. The methodology given by Jackson[1] for axisymmetric submarine-like forms is easy to implement and adequately accurate for initial estimates. By modifying the parameters characterising the vessel shape, it is useful to examine the impact of varying the form on the drag and power requirements.

AUVs typically operate at low speeds, i.e., low values of the Reynolds number (although beyond laminar flow); hence skin friction drag is expected to be more significant than form drag. Therefore, more attention should be paid to minimise the total exposed surface area (i.e., wetted surface area) of the vessel and any projections and appendages, rather than trying to optimise its overall form. A smaller vessel will have lower drag, but internal volume would be driven by considerations of stored energy and payload. For greater maximum speed, it is more efficient to increase power plant capacity rather than trying to reduce displacement.

Building Block Approach

Most of the internal volume of AUVs is devoted to energy storage, typically in the form of batteries. 'Payload' refers to the instruments, sensors or other

---

[1]   Jackson, 1992

objects that form the main purpose of deploying the AUV. The other major components are the thruster (usually integrated motor with propeller) and the electronics. The electronics include power distribution, mission computers, data storage media, data buses, etc. The typical 'blocks' of the internal volume of an AUV are depicted in Figure 4-1.

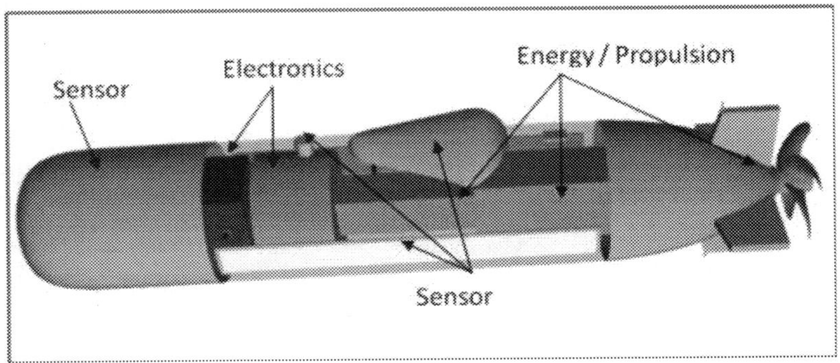

Figure 4-1. Blocks for equipment and systems in a typical AUV (Author)

The electronics are invariably located in a pressure-proof enclosure (called 'pressure hull' on submarines), which is cylindrical in shape for optimal strength against external hydrostatic pressure. Items located outside the pressure-proof enclosure are electrically connected to the internal electronics using pressure-resistant cables. All cables passing through the pressure hull require special 'penetrators' at the boundary to prevent ingress of sea water. An external fairing provides a streamlined shape enclosing both the pressure-proof enclosure as well as components outside the pressure hull exposed to sea water, as shown in Figure 4-2. This fairing typically has slots to allow free entry and exit of sea water as the vessel dives or surfaces.

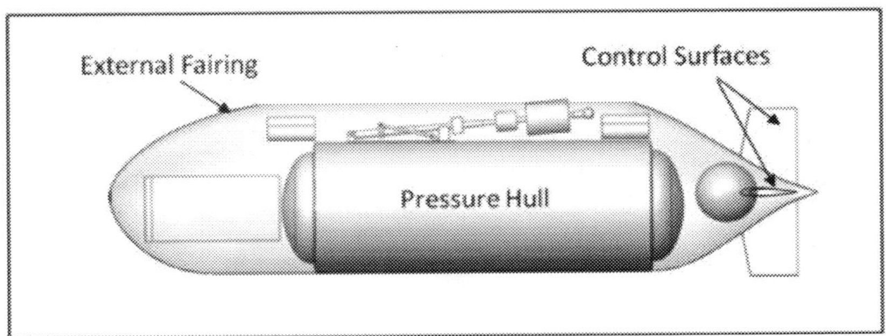

Figure 4-2. Pressure-hull enclosure and external fairing (Author)

Based on the requirements of payload, endurance, speed and other performance parameters, the initial estimate of internal volume can be made from the volume required for the major building blocks. These can then be refined considering a range of length and diameter values to conduct a parametric survey for correlating internal volume, size, form and speed/endurance.

Equipment- Driven Design

For small vehicles such as AUVs, the selection of one or a few major items of equipment such as type of battery, or multi-beam echosounder, or side scan sonar, could become a major factor affecting its size and form. These may also form the costliest as well as most significant components of the AUV (in terms of vehicle role or purpose). It would be prudent to start the design by identifying such major equipment and then design a cylindrical (or similar axisymmetric) enclosure to house them, as required for their functioning (Figure 4-3). From the viewpoint of functionality, this approach may result in a better estimate of vehicle weight, volume and form.

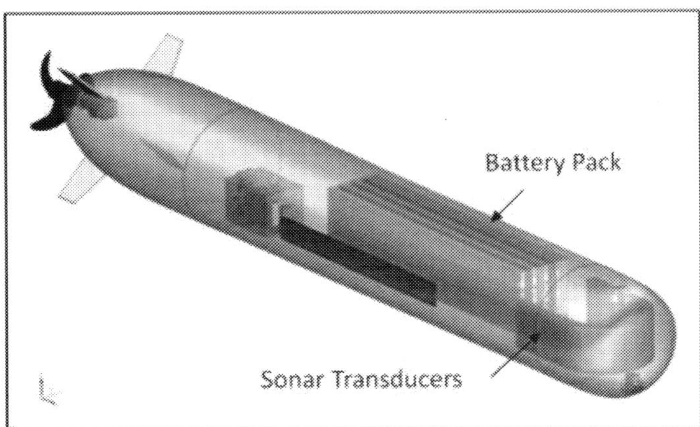

Figure 4-3. Examples of Major Equipment Determining AUV Design (Author)

Weight and Buoyancy Considerations

It is essential to remember the delicate balance of weight and buoyancy throughout the design of any vehicle that is intended to dive and surface. The weight and buoyancy need to be accurately estimated and matched for an

AUV (or submarine) to be able to dive and surface as intended. Imbalances are likely to occur due to inaccurate or incomplete data of vehicle components during design, variations during assembly or installation and due to variations in water density.

The major contributors to weight of a typical AUV are its batteries and its pressure-resistant structure. The greatest contributors to volume are the enclosed volume of the pressure hull and the volume of water displaced by large components outside the pressure hull (e.g., external pressure-compensated batteries).

Any imbalance between the weight and buoyancy of a vessel needs to be rectified either by addition of weights or by installation of (rigid) buoyancy foam. The former option is always preferable due to the relative ease of adding/removing weights, provided this is catered for in the design. Use of buoyancy foam material is a more difficult option to implement, since the shapes of their pieces must be customised and are not easily amenable to modification. Further, their compressibility needs to be evaluated carefully, particularly for very deep diving AUVs.

It is also essential that the centre of gravity and centre of buoyancy of the AUV, when neutrally buoyant in water, are in the same vertical line, thereby enabling the vehicle to maintain a horizontal attitude when it is not underway. The disposition of weights should also ensure that the VCG of the AUV is slightly below its VCB (by at least about 10 centimetres), for the AUV to remain upright in the submerged condition.

It is common practice in the AUV industry to design the vehicle to be slightly positively buoyant by default, so that in case of loss of power, the AUV would float up to the water surface. This approach means that the vehicle has to be driven downwards by application of control forces through its hydroplanes, which are effective only when the vehicle has forward speed.

Major Components and Systems of an AUV

Major physical components of an AUV include its structure, powering, energy storage, sensors, controls and computational modules (Figure 4-4).

In this section we discuss major design considerations for AUV sensors and payload, structural design, hull form, vehicle systems and software functions. In subsequent chapters, some of these systems would be covered in greater detail.

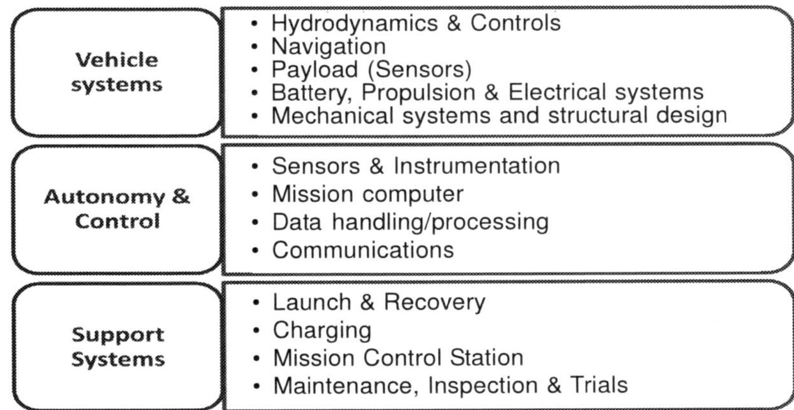

Figure 4-4. Salient systems and components in AUV design (Author)

Sensors and Payload

The 'payload' refers to the equipment for which the requirement of the UUV has been conceived in the first place. This includes equipment and components essential for sensing its environment and undertaking data collection, measurements, as well as safe deployment and operations. The vessel (UUV or ROV) must be designed around its mission or purpose, which translates to selection of payload first and then designing the vehicle around it.

The layout of sensors and payloads in the AUV designed by India's CSIR-CMERI is shown in Figure 4-5.

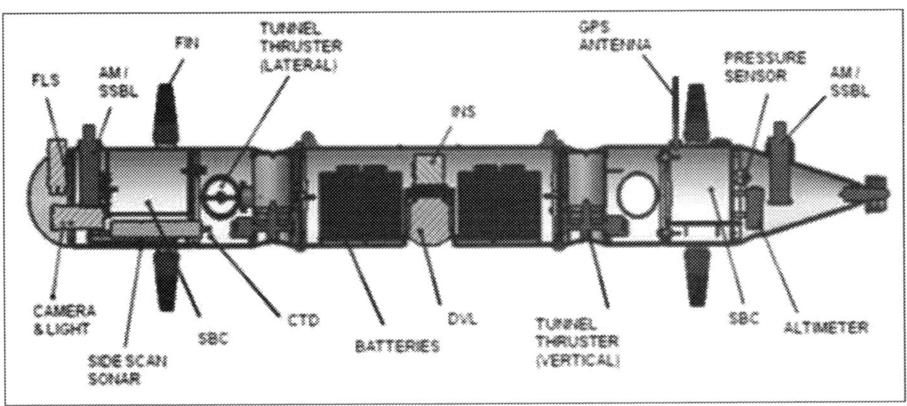

Figure 4-5. Components and Layout of AUV-150 developed by CMERI Durgapur, India[2]

---

[2]   Ambastha, et al., 2014

A comprehensive sensor suite is essential for a UUV to operate underwater for external data collection and reliable navigation. This would typically include:

- **Inertial Measurement Unit (IMU):** Provides acceleration and angular rate data for motion estimation.
- **Doppler Velocity Log (DVL):** Measures the vehicle's velocity relative to the seabed.
- **Pressure Sensors:** Measure depth and help with buoyancy control.
- **Sonar and Acoustic Sensors:** Facilitate obstacle detection, mapping, and communication.
- **Optical Sensors and Cameras:** Used for visual navigation, object recognition, and environmental monitoring.
- **Environmental Sensors:** Temperature, salinity, and current sensors help the vehicle adapt to changing conditions.

Some of the sensors are themselves examples of 'payload' in the sense that they carry out observations not only to support the operations of the UUV but also collect information that form the mission of the UUV. Thus, the payload could include environmental sensors for a scientific AUV, the optical sensors for a scientific or industrial AUV, and the acoustic sensors for a military AUV.

The choice of sensor suite on a particular AUV design is oriented towards its mission so that the requirements of 'payload' can be combined to the extent feasible with the instruments required for the navigation and operation of the vehicle itself. The sensors and layout of the Indian-design AUV *Maya* (by CSIR-NIO) is shown in Figure 4-6.

Structure, Materials and Hull Form

As discussed in the 'building block' and 'equipment-driven' approaches, the equipment (particularly the payload and batteries) to be fitted are the initial considerations for deciding the size of the pressure-resistant section(s) of the AUV.

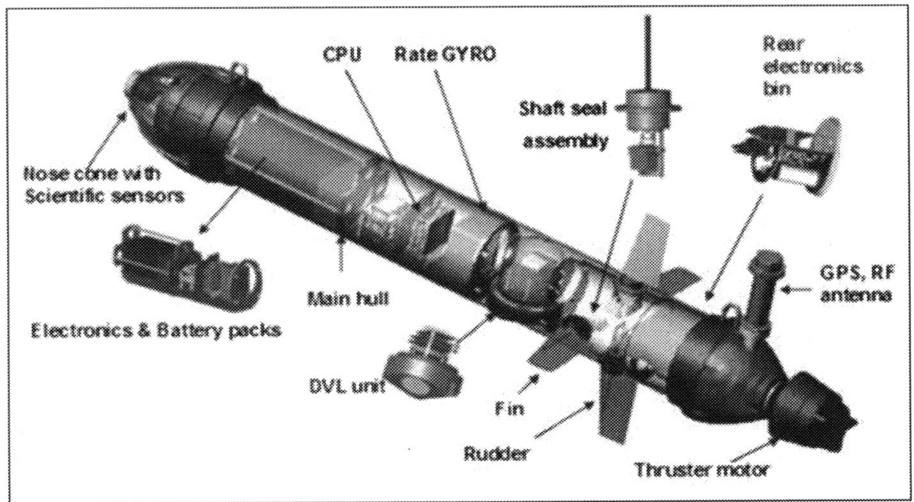

Figure 4-6. Components and Layout of AUV Maya developed by NIO Goa[3]

Structural Design

The structural design of the pressure-resistant portion is based on load due to external hydrostatic pressure at the maximum operating depth for which the AUV is to be designed. This structural load is so high that any other loads are rendered relatively insignificant for the design of the pressure hull. The shape selected is either a sphere, which is difficult to fabricate, or a cylinder with hemispherical (or 'dished') ends, which offers greater usable internal space and is more conducive to modular arrangement. The cylinder is stiffened by circular ring frames (longitudinal stiffening is not necessary).

Materials

The material of the pressure hull structure needs to have a high strength-to-weight ratio, display a suitable degree of elasticity, withstand corrosive environment and cyclic (fatigue) loading, and be feasible to fabricate.[4] Such metals include steel, aluminium or titanium alloy. Composite materials such as GRP or CFRP suffer from disadvantages of nonlinear behaviour under loading, and difficulty in inspection or repairs. Typical examples are given in Table 4-1.

---

[3]   Madhan, et al., 2007

[4]   Fathallah & Helal, 2016

Table 4-1. High-Strength Materials for Underwater Pressure Hull Structures[5]

Material	Specific Density (g/cm³)	Young's Modulus (GPa)	Compressive Yield Strength (MPa)	Ultimate Strength (MPa)	Poisson Ratio
Aluminium (6061)	2.700	70	276	310	0.33
Stainless Steel (316)	8.000	195	290	550	0.30
HY80 Steel	7.828	210	550	690	0.29
HY100 Steel	7.828	210	690	793.5	0.29
Titanium alloy (Ti-6Al-4V)	4.430	110	827	1000	0.30
ABS (moulded)	1.050	2.2	65	40	0.35
Polycarbonate (extruded)	1.200	2.4	70	67	0.37

Materials commonly used for various structural components (pressure hull and otherwise) in underwater vehicles are listed in Table 4-2.

Table 4-2. Commonly used structural materials in underwater vehicles[6]

Material Type	Material	Used in Frames	Used in Pressure Hull
Metals	Steel	Yes	Yes
	Aluminium	Yes	Yes
	Titanium	Yes	Yes
Plastics	Acrylic	No	Yes (Viewports)
	ABS, PVC	Yes (Brackets, parts)	Yes (shallow water)
Composites	GRP	Yes (Fairings)	Experimental
	Syntactic Foam	Floatation	No
Other	Glass	No	Experimental
	Ceramic	No	Experimental
	Rubber	Bumpers	O-Ring Seals
	Lead	Yes (Ballast)	No

Hull Form

The external hull shape of an AUV should be streamlined for minimising resistance (i.e., drag) of motion through water. More importantly, the external form is decided based upon the diameter and length of the pressure-proof enclosures that it encloses.

---

[5] Allmendinger, 1990; Moore, Bohm, & Jensen, 2010.

[6] Moore, Bohm, & Jensen, 2010

AUVs typically operate at low speeds (3 to 5 knots, i.e., about 1.5 to 2.5 m/s). Low speed is preferred for better performance of sensors and for minimising power consumption since power requirement increases steeply with speed (approximately proportional to the cube of speed).

At such low speeds, it is the frictional drag component that dominates for which it is more beneficial to minimise the external surface area of the AUV rather than refining its shape. A cylindrical hull with fairings at both ends is the typical form selected for AUVs, and its drag coefficient can initially be estimated from empirical data accordingly.

Other hull forms (apart from cylindrical axisymmetric type) have also been adopted and utilised for specific applications, such as 'flatfish', 'blended wing' or wing-appended forms.

Hydrodynamics and Control

The hydrodynamic shape of the bare hull needs to be made controllable and manoeuvrable. This is commonly achieved by control surfaces (fins) at the stern (rear) of the AUV, which can be deflected by actuators to achieve turning in the horizontal plane, and pitching up or down in the vertical plane. The sizes of the control surfaces and their shapes (profiles) can be estimated using empirical data to start with.

The mathematical model of motion of the vehicle correlates its form (shape) with the location and shape of its control surfaces, speed, weight (and CG), the control forces (deflection of fins) with the resultant orientation and trajectory of the AUV in the three-dimensional underwater domain. Here, the vehicle characteristics are represented by its mass matrix and (non-dimensional) hydrodynamic coefficients, while the applied forces are both external and due to control surfaces deflection and speed. Solving these six differential equations of motion for finite time steps enables the trajectory to be estimated as a function of time.

The mathematical model of an AUV's motion is therefore a set of nonlinear dynamic equations representing the vehicle's rigid-body and hydrodynamic behaviour.[7] This model describes how the vehicle responds to control inputs and environmental forces, using physical laws (Newton/Euler equations) adapted for underwater dynamics, including effects like buoyancy, drag, added

---

[7]   Ray, 2016

mass, and Coriolis forces. This serves as the essential basis for designing its Guidance, Navigation, and Control (GNC) systems,[8] as follows:

- **Guidance:** The Guidance system is responsible for generating the trajectory (path in underwater space) for the vehicle to follow. When waypoints are defined, it generates a trajectory to feed the controllers set points. The mathematical model predicts how different sequences of control inputs will change the AUV's trajectory and orientation; path planners and trajectory generators use the model to ensure feasible and safe motion.[9]

- **Navigation:** State estimators (e.g., Kalman filters) combine model-predicted states with real-time sensor measurements to estimate the true position, velocity, and attitude of the AUV, accounting for uncertainty and disturbances.

- **Control:** The control laws (such as PID or model-based controllers) are derived using simplified linearised or full nonlinear models of the vehicle, ensuring stability and desired tracking behaviour under anticipated environmental and actuation conditions.

Model-based GNC design cycles through model simulation, controller design, and experimental validation—each stage refining both the mathematical model (to better match real dynamics) and the GNC architecture for precision, robustness, and safety. Thus, the mathematical model is the foundation that structurally links AUV physics to every facet of motion planning, perception, and actuation.[10] Design of motion control and guidance has been reported for various Indian AUVs as well.[11]

Vehicle Systems

The physical architecture of an autonomous underwater vehicle typically consists of several key subsystems, described below. Some of these will be covered in further detail in respective chapters subsequently.

---

8   Koskinen, 2020

9   Maurya, 2015

10   Valeriano-Medina, 2012

11   Maurya, 2015; Upadhyay, 2015; Das, 2010.

Energy and Power Systems

The energy and power requirements of an AUV are the main drivers for its size. The speed requirement and size, in turn, govern the power requirements, thus forming a tight conundrum.

Batteries are the primary source of energy, with popular choices being lithium-ion secondary cells (to be recharged ashore or on a mother ship). Emerging technologies that are being explored include fuel cells, super-capacitors, and hybrid power systems.

Efficient power management systems are needed to balance the energy consumption between propulsion, computing, and sensor operation.

Actuation Systems

- **Thrusters:** Electric thrusters (brushless DC motors) are the primary means of propulsion. They are controlled via electronic speed controllers (ESCs) and can be arranged in various configurations (e.g., vectored arrays) to provide full six-degrees-of-freedom (6-DOF) manoeuvrability.

- **Control Surfaces:** For vehicles that use hydrodynamic control (such as gliders), movable fins, rudders, or internal ballast adjustment mechanisms are crucial for steering and maintaining stability.

Onboard Computing and Communication

- **Central Processing Unit (CPU)/Embedded Systems:** Modern AUVs employ ruggedized embedded computers (often running Linux or real-time operating systems) that handle mission planning, sensor data processing, and control algorithms.

- **Communication Modules:** Due to the limited range of radio signals underwater, AUVs often rely on acoustic modems for underwater communication and satellite or cellular links for surface data transmission.

- **Data Storage:** Onboard data storage systems are used to log sensor data and mission parameters.

Software Functions

Autonomous operations of an underwater vehicle require robust, fault-tolerant and capable software, linked to reliable sensors, actuators and components. These aspects become crucial for safe operations of an AUV since it is normally not in communication with a shore control station and only intermittently comes to surface for transmitting/receiving updates.

The various functions of software in an autonomous underwater vehicle are outlined below.

Mission Planning and Decision Making

- **High-Level Autonomy:** Software frameworks allow the operator to define missions using waypoints, tasks, and environmental constraints. Advanced planning modules incorporate AI and machine learning algorithms to enable real-time re-planning in response to dynamic conditions.

- **Behaviour-Based Control:** A layered architecture often separates strategic decision-making from low-level control, where the high-level planner issues commands that are translated into specific control actions.

Real-Time Control and Data Fusion

- **Control Loops:** Real-time control algorithms (e.g., PID, model predictive control, sliding mode control) ensure that the AUV maintains its trajectory, speed, and orientation.

- **Sensor Fusion:** Data from multiple sensors (IMUs, DVLs, pressure sensors, sonars, GPS when surfaced) are fused to produce a robust estimate of the vehicle's state. Techniques such as Kalman filtering and particle filtering are commonly used.

Diagnostic and Fault Tolerance

- **Health Monitoring:** Software routines continuously monitor the status of hardware components and sensor readings, triggering safe modes or re-planning if anomalies are detected.

- **Redundancy Management:** In high-autonomy AUVs, redundant hardware and software pathways ensure that a failure in one subsystem does not lead to mission abort.

Systems Engineering Approach

The Systems Engineering approach, initially formalised for aerospace and defence applications, is a standardised methodology of progressing from user requirements to analysis and testing, with documentation and reviews at various stages.[12] Systems engineering (SE) provides a structured framework for designing, developing, and integrating complex systems.[13] For AUVs, SE methods are essential due to the multidisciplinary nature of the system and the need to satisfy strict operational requirements. This section discusses key SE methodologies applied to AUV development.

The Vee Model

The Vee model is a widely used representation in systems engineering that emphasises both the decomposition of requirements and the integration and verification of subsystems. The model's left side represents the progressive decomposition of system requirements and the design process, while the right side represents system integration and validation.

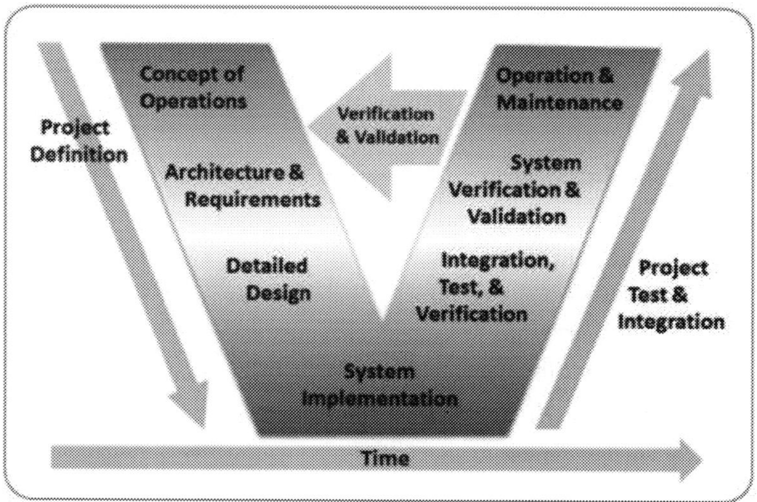

Figure 4-7. Graphical 'Vee' Model in Systems Engineering
(Source: INCOSE[14])

---

[12]   Dieter, 2021

[13]   NASA, 2017

[14]   INCOSE, 2015

- **Requirements Definition:** At the apex of the left side, stakeholders define mission requirements, operational constraints, and performance objectives.

- **Functional Decomposition:** These requirements are then decomposed into functional and technical specifications that guide the design of both hardware and software subsystems.

- **Design and Implementation:** As the model descends, detailed designs are created for individual modules (both mechanical and electronic) and software components.

- **Integration and Verification:** The right side of the Vee model represents the integration of subsystems and their verification against the original requirements.

- **Validation and Testing:** The final step involves validating the complete system in realistic conditions, such as field trials or simulated environments.

By following the Vee model, AUV developers can ensure that every design decision is traceable back to user requirements, and that each module is verified and validated before final integration. This iterative process helps reduce errors and improves overall system reliability.

Requirements Mapping

A fundamental aspect of systems engineering is the mapping of user requirements to system functions. Tools such as the House of Quality (HoQ) and Quality Function Deployment (QFD) are often used to translate stakeholder needs into engineering specifications. In AUV development, this process involves the following specific activities:

- **Identifying Core Functions:** Determining the essential functions an AUV must perform—such as propulsion, sensing, navigation, communication, and power management.

- **Mapping to Subsystems:** Dividing these functions among modular hardware and software components. For example, the navigation function might be implemented by a combination of inertial measurement units (IMUs), GPS (when surfaced), and acoustic positioning systems.

- **Establishing Interfaces:** Defining standard interfaces and data protocols that enable communication between these subsystems. Standardization ensures that modules developed by different teams can be integrated without extensive rework.

Integration, Verification, and Validation (IV&V)

IV&V are critical to systems engineering and are applied at multiple levels in AUV development:

- **Unit Testing:** Each hardware and software module is tested independently to verify that it meets its specifications.

- **Subsystem Integration Testing:** Modules are integrated into larger subsystems, and tests are conducted to verify that the interfaces and interactions work as expected.

- **System-Level Testing:** The complete AUV is tested—both in laboratory conditions (e.g., water tank tests) and in real operational environments (e.g., sea trials). This stage verifies that all components work together to fulfil the overall mission.

- **Fault Injection and Recovery Testing:** To assess fault tolerance, systems are subjected to simulated faults (e.g., sensor failure, communication dropouts) and observed for their ability to recover or degrade gracefully.

Through rigorous IV&V processes, systems engineers ensure that the AUV not only meets design requirements but is robust enough to handle the uncertainties of the underwater environment.

Hardware and Software

Modular design in both hardware and software provides flexibility to adapt AUVs to varied and challenging missions while ensuring that each component can be developed, verified, and upgraded independently. Mission computer modularity and standardised interfaces further streamline the integration process, allowing diverse subsystems to interoperate effectively.

Layered Architecture

The hardware and software of an AUV are typically organized into layers, such as a low-level device layer, an instrument/control layer, a mission management layer, and a user interface layer. This separation isolates each functionality and allows independent development and testing.[15]

Typical AUV hardware and software architecture includes the following layers:

- **Device and Hardware Abstraction Layer:**
  - o Provides drivers and interfaces to sensors, actuators, and communication hardware.
  - o Abstracts the low-level hardware details so that higher layers can operate without concern for specifics such as serial protocols or analog-to-digital conversion.

- **Middleware/Communication Layer:**
  - o Implements standard data exchange protocols (e.g., publish/subscribe, remote procedure calls).
  - o Often built on open standards (such as DDS or ROS topics) to facilitate interoperability and distributed processing.[16]
  - o Provides a uniform messaging interface to support modularity and reusability.

- **Control and Decision Layer (Backseat Driver):**
  - o Contains high-level control algorithms, mission planning, and decision-making logic.
  - o Utilizes models such as behaviour trees or state machines to structure complex autonomous behaviour.
  - o Implements fault tolerance measures by monitoring subsystem status and re-planning missions if necessary.

- **Application/User Interface Layer:**
  - o Provides interfaces for mission configuration, monitoring, and control by human operators.
  - o Often includes visualization tools and diagnostic displays.
  - o Facilitates user interaction through standard graphical user interfaces (GUIs) or web-based dashboards.

---

[15] AUTOSAR

[16] ROS

- **Data Management and Logging Layer:**
  - o Manages data storage, retrieval, and synchronization between onboard systems and surface stations.
  - o Uses standardized data protocols (e.g., JSON, XML, or binary formats) to ensure that data can be processed by third-party tools.
  - o Implements real-time logging and fault reporting mechanisms.

Fault Tolerance Strategies

Fault tolerance is essential for AUVs, given the difficulty and expense of retrieving a vehicle that has failed underwater. The software architecture includes mechanisms for detecting and recovering from faults. Redundant modules, graceful degradation, and exception handling strategies ensure that critical functions remain operational even when some components fail. Strategies for achieving fault tolerance include the following:

- **Redundancy:** Incorporating duplicate sensors, communication links, and processing units so that the failure of one component does not compromise the entire system. For example, a redundant thruster design can ensure that if one thruster fails, the vehicle can still maintain control.

- **Graceful Degradation:** Designing systems to reduce performance gradually rather than failing catastrophically. If a non-critical module fails, the system can still operate in a degraded mode until recovery is possible.

- **Self-Diagnostics and Reconfiguration:** Implementing health monitoring algorithms that continuously assess the status of each module. When a fault is detected, the system can reconfigure itself (e.g., by switching to backup modules or rerouting data flows) to maintain safe operation.

- **Modular Isolation:** Using techniques such as process isolation and sandboxing to prevent a fault in one software module from propagating to others.

- **Robust Communication Protocols:** Ensuring that data protocols include error detection, correction, and retransmission mechanisms to handle transient communication failures.

Standardised Interfaces and Data Protocols

In complex systems like AUVs, standardization plays a crucial role in ensuring interoperability and ease of integration. Systems engineering promotes the use of standardised interfaces and data protocols through the use of the following features:

- Standard Communication Protocols
- Interface Definition Languages (IDLs)
- Electrical and Mechanical Standards
- Data Abstraction Layers

Modules communicate using standard protocols (e.g., publish/subscribe models, remote procedure calls) and common data formats (such as JSON, XML, or custom binary protocols) that ensure interoperability between software components. By adhering to these standards, developers reduce integration complexity and improve system maintainability.

Conclusion

This chapter has outlined the approaches for embarking upon modern AUV design. The major components and systems of AUVs and their salient design considerations have been outlined. Typical architecture, payload, layout, design considerations for structures and hull form, and major systems have been described that would prepare a design team of engineers to approach their task with confidence.

Applying systems engineering methodologies, such as the Vee model and iterative IV&V, ensures that design decisions are fully traceable to user requirements and that every subsystem is rigorously tested before final integration.

Design and development of AUVs require careful balance of hardware innovation, software robustness, and rigorous systems engineering. By integrating robust Systems Engineering methodologies with modular design philosophy, the next generation of AUVs will be better equipped to perform high-risk, long-duration missions in unpredictable underwater environments.

With this overview of all components, systems and design considerations of an AUV, we will examine several of these aspects in greater detail in respective chapters that follow. Before we do that, in the next chapter, we touch upon standards relevant for AUV design, development and testing.

References

Allmendinger, E. (1990). *Submersible Vehicle Systems Design.* Jersey City: Society of Naval Architects and Marine Engineers.

Ambastha, S., Das, S., Pal, D., Nandy, S., Shome, S., & Banerjee, S. (2014). Underwater terrain mapping with a 5-DOF AUV. *Indian Journal of Geo-Marine Sciences,* 43, 106-110. Retrieved May 2025 from https://www.researchgate.net/publication/271079407_Underwater_terrain_mapping_with_a_5-DOF_AUV

Aristizábal, L., Zuluaga, C., Rúa, S., & Vásquez, R. (2021, May 11). Modular Hardware Architecture for the Development of Underwater Vehicles Based on Systems Engineering. *Journal of Marine Science and Engineering,* 9, 516. doi:10.3390/jmse9050516

AUTOSAR. (n.d.). AUTomotive Open System ARchitecture. *AUTomotive Open System ARchitecture.*

Das, S.K., Pal, D., Nandy, S., Kumar, V., Shome, S.N., Mahanti, B. (2010). Control Architecture for AUV-150: A Systems Approach. In: Vadakkepat, P., et al. Trends in Intelligent Robotics. FIRA 2010. *Communications in Computer and Information Science,* 103. Springer, Berlin, Heidelberg. https://doi.org/10.1007/978-3-642-15810-0_6

Dieter, G., & Schmidt, L. (2021). Engineering Design. *Engineering Design.* McGraw-Hill Education.

Fathallah, E., & Helal, M. (April 2016). Optimum Structural Design of Deep Submarine Pressure hull to achieve Minimum Weight. *11th International Conference on Civil and Architecture Engineering.* Cairo.

Hwang, J., Bose, N., Millar, G., Gillard, A., Nguyen, H., & Williams, G. (2014). Enhancement of AUV Autonomy Using Backseat Driver Control Architecture. *International Journal of Mechanical Engineering and Robotics Research,* 10, 292-300. doi:10.18178/ijmerr.10.6.292-300

INCOSE (2015). In *INCOSE Systems Engineering Handbook: A Guide for System Life Cycle Processes and Activities.* Wiley.

Jackson, H. A. (1992). Fundamentals of Submarine Concept Design. *SNAME Transactions,* 100, 419-448.

Koskinen, K. (2020). Model-based Guidance, Navigation and Control architecture for an Autonomous Underwater Vehicle. Global Oceans 2020: Singapore – U.S. Gulf Coast, IEEE.

Macenski, S., Foote, T., Gerkey, B., Lalancette, C., Woodall, W. (2022, May). Robot Operating System 2: Design, architecture, and uses in the wild. *Science Robotics,*7.

Madhan, R., Desa, E., Prabhudesai, S., Sebastiao, L., Pascoal, A., Mascarenhas, A., & Khalap, S. (2007). Mechanical design and development aspects of a small AUV—Maya. *IFAC Conference on Manoeuvring and Control of Marine Craft.* Retrieved May 2025, from http://drs.nio.org/drs/handle/2264/702

Maurya, P., Desa, E., Pascoal, A., Barros, E. et al (20065). Control of the Maya AUV in the vertical and horizontal planes: Theory and practical results. In: Proceedings of 7th IFAC Conference on Manoeuvring and Control of Marine Craft. Instituto Superior Técnico. Lisbon, Portugal. Retrieved from: https://isr.tecnico.ulisboa.pt/wp-content/uploads/2015/05/ 1531_C4.pdf

Moore, S. W., Bohm, H., & Jensen, V. (2010). *Underwater Robotics: Science, Design and Fabrication.* Monterey, CA: Marine Advanced Technology Education Centre.

MOOS-IvP. (n.d.). Retrieved from https://www.moos-ivp.org.

NASA. (2017). *NASA Systems Engineering Handbook,* USA.

Ray, A., Sen, D., Singh, S.N., Seshadri, V. (2016, September). Manoeuvrability Assessment of Underwater Vehicles by Semi-Empirical, Numerical and Experimental Studies. 31st Symposium on Naval Hydrodynamics, Office of Naval Research, Monterey, CA, USA.

ROS. (n.d.). Retrieved from https://www.ros.org.

Ullman, D. (2017). In *The Mechanical Design Process.* David Ullman LLC.

Upadhyay, V., Gupta, S., Dubey, A.C., et al., Design and motion control of Autonomous Underwater Vehicle, Amogh (2015). IEEE Underwater Technology, Chennai, India, pp. 1-9, doi: 10.1109/UT.2015.7108287.

Valeriano-Medina, Y., Martínez, A., Hernández, L., Sahli, H., Rodríguez, Y., & Cañizares, J. R. (2012). Dynamic model for an autonomous underwater vehicle based on experimental data. *Mathematical and Computer Modelling of Dynamical Systems,* 19(2), 175–200. https://doi.org/10.1080/13873954.2012.717226

Zuluaga, C., Aristizábal, L., Rúa, S., Franco, D., Osorio, D., & Vásquez, R. (2022, March 25). Development of a Modular Software Architecture for Underwater Vehicles Using Systems Engineering. *Journal of Marine Science and Engineering,* 10, 464. doi:10.3390/jmse10040464

5

Standards

Given the complexity and criticality of underwater submersibles and UUVs, certification and adherence to rigorous design rules are highly recommended. This chapter provides an overview of standards and rules for underwater submersibles and vehicles. We start with role of classification societies and standards for marine structures, materials and components. Thereafter, standards for interface and data protocols, autonomy, software architecture are discussed, which are crucial for seamless integration, communication and data transfer.

Role of Classification Societies

Classification societies have historically been the backbone of maritime safety. Their primary role is to set and enforce standards for the design, construction, and periodic survey of ships and marine structures. Although initially focused on surface vessels, many classification societies now publish detailed rules for underwater vehicles and submersibles.

Certification of underwater vehicles is typically handled by classification societies (ABS, DNV, ClassNK, LR) and national regulatory bodies. These organizations provide third-party verification that a vessel meets international and local safety standards. Certification involves rigorous design reviews, material testing, and periodic surveys to ensure continued compliance throughout the vessel's service life.

Certification agencies have decades of experience and a well-established framework for assessing vessel integrity and safety. They cover a wide range of parameters from structural design to electrical systems.

However, many existing standards were developed for manned vessels or traditional maritime structures and may not fully address the unique challenges of modern unmanned and autonomous systems. For instance, rapid software updates, adaptive autonomous functions, and network-centric operations are areas that require more dynamic evaluation than traditional static certification procedures provide. Further, certification involves substantial cost.

The scope of standards and coverage provided by various Classification Societies for underwater vehicles are discussed further.

Lloyd's Register

Lloyd's Register (LR) is the world's oldest Classification Society, with headquarters in London. It has expanded its scope to include unmanned and autonomous systems.

The LR Code for Unmanned Marine Systems provides requirements for the design, construction, and operation of unmanned marine systems, including underwater vehicles. It covers aspects such as system reliability, fault tolerance, and environmental protection.[1]

LR emphasizes risk management throughout the vehicle's lifecycle and offers certification services that verify compliance with LR standards. These services provide additional assurance that unmanned systems meet the highest safety standards.

American Bureau of Shipping

The American Bureau of Shipping (ABS) has issued comprehensive rules for underwater vehicles, including submersibles and unmanned systems,[2] covering the following aspects:

- **Structural Design and Stability:** Requirements for hull integrity, material selection, and stability calculations under high hydrostatic pressures.
- **Power and Propulsion:** Standards for the design and installation of power systems and propulsion components to ensure reliable operation in deep water.

---

[1] LR, 2020

[2] ABS Rules for Building and Classing Underwater Vehicles, Systems and Hyperbaric Facilities, 2025

- **Electrical and Control Systems:** Guidelines for the safe installation and operation of electrical systems, including redundant control architectures and fault tolerance.
- **Survey and Inspection:** Procedures for periodic surveys and testing to ensure continued compliance with safety standards.

ABS rules are updated periodically to reflect new technologies and operational experience. For example, ABS's 'Requirements for Autonomous and Remote Control Functions' provide detailed criteria that designers must meet to ensure that AUVs and ROVs can operate safely at depth.[3]

Det Norske Veritas

Det Norske Veritas (DNV) is another key organisation for maritime classification and certification, which merged in 2013 with the erstwhile Germanischer Lloyd (GL), which had its own set of rules for underwater vehicles. DNV has a suite of standards and guidelines for unmanned underwater vehicles:

- **DNVGL-RU-UWT:** This standard addresses the classification of unmanned underwater vehicles and provides requirements for design, manufacture, and operation.[4]
- **Risk and Reliability Assessment:** DNV guidelines emphasize probabilistic risk assessments, considering both hardware and software reliability. These assessments help to quantify the probability of failure and to implement redundant systems as needed.
- **Operational Qualification:** DNV also provides certification schemes that cover not only the design and construction of UUVs but also their operational performance. This includes specific testing protocols for endurance, sensor accuracy, and environmental resilience.

ClassNK

ClassNK (Nippon Kaiji Kyokai) is Japan's classification society. It has developed guidelines for underwater vehicles (ROVs and AUVs), covering aspects from

---

[3]   ABS Requirements for Autonomous and Remote Control Functions, 2024

[4]   DNV, Rules for Classification: Underwater Technology — DNVGL-RU-UWT Pt.3 Ch.2, 2015

structural design to electronic systems and software control. These aim to ensure safety, performance, and reliable operation in marine environments.[5]

Standards for Structures, Materials, Components

Marine submersibles and underwater vehicles must be constructed from materials that can withstand high pressures, corrosive salt water, and dynamic loads. ASME (American Society of Mechanical Engineers) and related standards play a crucial role in this domain.

Pressure Vessel Construction

ASME Boiler and Pressure Vessel Code provides guidelines for the construction of pressure vessels, whose section VIII is one of the primary standards referenced by designers of deep-sea submersibles and unmanned vehicles.[6] This covers the following aspects:

- **Design Requirements:** It outlines requirements for design, fabrication, inspection, and testing of pressure vessels that are used in underwater vehicles.

- **Material Selection:** This section covers the selection of materials that have adequate strength, ductility, and resistance to corrosion under high-pressure conditions.

- **Welding and Fabrication Procedures:** It specifies approved welding techniques and quality assurance processes to ensure that pressure vessels remain structurally sound throughout their service life.

For submersibles that carry human occupants, ASME PVHO-1 provides rigorous safety criteria ensuring that pressure vessels are designed to maintain structural integrity and safe operating conditions even under emergency scenarios. Compliance with ASME PVHO-1 is often a prerequisite for certification by maritime authorities, ensuring that submersibles meet or exceed safety expectations.

Marine Materials and Corrosion

In addition to pressure vessel codes, there are several standards related to marine materials:

---

5   ClassNK, 2021
6   ASME Boiler and Pressure Vessel Code - Section VIII, 2019

- **ASTM Standards:** The American Society for Testing and Materials (ASTM) has numerous standards governing the properties of marine materials, including corrosion resistance, mechanical properties, and environmental degradation. For example, ASTM A995/A995M covers structural steel castings for marine applications.
- **ISO Standards:** The International Organisation for Standardisation (ISO) publishes standards such as ISO 12944, which deals with corrosion protection of steel structures by protective paint systems—a key consideration for underwater vehicles.[7]
- **DNV-OS-F101:** Developed by Det Norske Veritas, this standard specifies requirements for the design, fabrication, and operation of offshore structures, including underwater vehicles, with a focus on material performance and corrosion protection.

These standards help ensure that underwater vehicles are constructed from materials that can endure the rigors of long-term submersion and extreme conditions.

Underwater Components

Underwater connectors, cables, and related components are critical for ensuring reliable electrical and data connections in harsh marine environments. Several international and military standards govern these components:

Both military and industrial UUVs incorporate connectors that have been certified to these standards, providing a high level of confidence in their long-term performance under operational conditions. Reliable underwater connectors are vital for maintaining electrical integrity and data communication in unmanned systems.

MIL Standards for Equipment

MIL-STD-810 is a US military standard that defines environmental engineering considerations and laboratory tests for military equipment.[8] The standard includes tests for temperature extremes, water immersion, vibration, shock, and other environmental factors. Underwater connectors and cables

---

[7]   ISO 12944: Corrosion Protection of Steel Structures by Protective Paint Systems, 2017

[8]   MIL-STD-810H: Test Method Standard: Environmental Engineering Considerations And Laboratory Tests, 2019

must meet the specified requirements to ensure that they maintain functionality in rugged, harsh environments.

MIL-DTL-38999 is a US military standard for electrical connectors used in demanding environments. It specifies the design, material, and performance requirements for connectors that must operate reliably underwater.[9]

IP Code for Connectors and Enclosures

IEC 60529 (IP code) is widely adopted in the design and testing of marine connectors and electronic enclosures, ensuring consistent quality and performance. This defines the degrees of protection provided by enclosures.[10] Underwater connectors, enclosures, and cables are rated according to their ability to resist water ingress and particulate matter. For underwater applications, an IP68 rating (or higher) is typically required to ensure continuous operation in submerged conditions. Many connectors and enclosures used in underwater vehicles are designed to meet or exceed IP68 or IP69K ratings as defined by IEC 60529. This ensures that critical electronics remain protected from water ingress.

Examples of Applications

To illustrate how these standards and certification requirements are applied in practice, a few examples are presented here.

Case Study: ABS and REMUS 6000

The REMUS 6000, widely used in both scientific and military applications, is designed in accordance with ABS rules. Key aspects include:

- **Structural Integrity:** The hull design adheres to ABS guidelines for deep-water operations, ensuring that the vehicle can withstand pressures at depths of up to 6,000 metres.

- **Electrical and Control Systems:** ABS rules ensure that the electronic and control systems are designed for redundancy and fault tolerance. Regular surveys and inspections by ABS-certified surveyors validate the ongoing compliance of the vehicle.

---

9   MIL-DTL-38999L: Detail Specification: Connectors, 2008
10   IEC 60529: Degrees of Protection Provided by Enclosures (IP Code), 2013

- **Operational Certification:** REMUS 6000 underwent extensive testing and certification by ABS before being deployed operationally by the US Navy. This certification covers design, construction, and in-service performance.

Case Study: DNV and Industrial UUVs

DNV certification plays a crucial role in the industrial deployment of UUVs:

- **Pipeline Inspection UUVs:** UUVs used for pipeline and cable inspections are designed to meet DNVGL-ST-0140 standards.[11] These vehicles are required to operate reliably in corrosive, high-pressure environments typical of offshore oil fields.

- **Risk and Reliability:** DNV guidelines incorporate rigorous risk assessment and probabilistic failure analysis. UUVs certified by DNV have undergone extensive tests that simulate operational conditions to ensure their reliability.

- **Integration with Operational Practices:** DNV-certified UUVs are integrated into routine industrial inspections, where certification ensures that inspections are performed safely and that the vehicles can be relied upon to detect anomalies or defects.

Case Study: ASME in Submersible Design

For manned submersibles, ASME standards are critical:

- **Pressure Vessel Design:** ASME section VIII governs the construction of pressure vessels used in submersibles. Compliance with these standards is mandatory for human-occupied underwater vehicles.

- **Welding and Material Specifications:** The ASME standards dictate stringent procedures for welding and material selection to ensure that the pressure hull can safely contain the internal atmosphere at great depths.

- **Operational Safety:** Submersibles designed for research or tourism (such as those used in deep-sea exploration) must adhere to ASME PVHO-1 standards, ensuring that all components have been rigorously tested for safety and durability.

---

[11] DNV, Rules for Classification: Underwater Technology — DNVGL-RU-UWT Pt.4 Ch.7, 2018

Standards for Interfaces and Data Protocols

UUVs are essentially nodes in a network of sensor nodes for data collection about the underwater domain. Further, each UUV incorporates several sensors that need to interface and communicate with the mission computer of the vehicle to enable effective data fusion.

Therefore, standardisation of interfaces and data communication protocols becomes imperative for integrating a range of sensors within a UUV, as well as for integrating a number of UUVs in a sensor network for coordinated operations.

Ideally, the sensor nodes underwater should complement any unmanned sensor nodes in the air, and suitable data fusion should be possible in near-real-time at shore control stations. From the viewpoint of multi-domain operations of unmanned systems, the conceptual requirements of C4 (Command, Control, Communication and Computers) have been represented by the US DoD as shown in Figure 5-1.

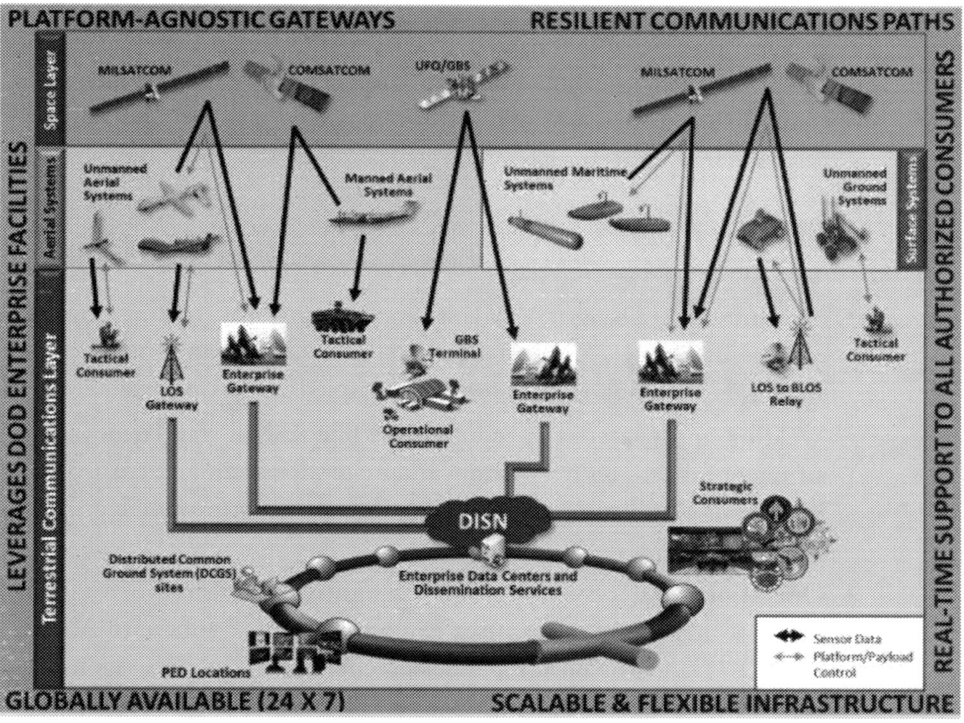

Figure 5-1. High-Level C4 Operational Concept
(Source: US DoD[12])

---

[12] US Department of Defense, 2013

Approach to Standardisation

Standardisation to enable compatibility between components, interfaces and data exchange can be addressed for AUVs and ROVs by the following means:

- **Defining Communication Protocols:** Adopting common protocols for data exchange (such as DDS, MQTT, or ROS topics) that specify message formats, timing, and quality-of-service (QoS) parameters.

- **Interface Standardisation:** Specifying electrical and mechanical interfaces for hardware modules (e.g., connectors, power specifications, and signalling standards) to guarantee compatibility. For instance, using standardized connectors (such as MIL-SPEC wet-mate-able connectors) and communication buses (such as CAN or Ethernet) facilitates integration.

- **Data Abstraction:** Using middleware that abstracts raw data into standardized structures (e.g., using a common data model for sensor outputs). This allows software modules to process data uniformly, regardless of the sensor manufacturer.

- **Use of Open Standards:** Leveraging open-source platforms (e.g., the Robot Operating System (ROS)) which provide well-documented standards for message passing, service definitions, and system configuration. Open standards ensure that new components are easier to integrate and that the overall system remains scalable and maintainable.

These aspects are summarised in Table 5-1 with examples of typical relevant protocols and standards.

Standardising Autonomy Architecture

The architecture for autonomous operation of the Mission Computer of an AUV needs to consider the connectivity and data exchange between components of vehicle, to ensure health monitoring and effective decision making in planned as well as unplanned scenarios.

The US Navy has developed the Unmanned Maritime Autonomy Architecture (UMAA) to define a comprehensive data model and a standardized group of services required of maritime vessels and systems. It covers vehicle-based autonomy architecture for unmanned maritime platforms, as represented for a typical Unmanned Marine Vehicle (UMV) in Figure 5-2.

Table 5-1. Overview of Interfaces and Data Protocols in AUV Software

Layer/ Interface	Typical Protocol/ Standard	Description	Advantages
Hardware Abstraction	RS-232/RS-422, I²C, SPI	Standard electrical communication for sensors, actuators	Low-level simplicity; well-established hardware interfaces
Middleware	ROS, MOOS-IvP, DDS	Publish/subscribe messaging; standardized APIs for data exchange	Decouples modules; supports distributed processing
Data Formats	JSON, XML, Protocol Buffers	Common formats for representing sensor data and messages	Facilitates interoperability; human-readable if needed
Networking	Ethernet, CAN Bus	High-speed, reliable data exchange between modules	Supports high data rates; robust against interference
Real-Time communication	SOME/IP (Not common underwater), DDS QoS	Ensures timely delivery of control commands and sensor data	Guarantees low latency; supports critical mission functions
Acoustic Modem communication	Micromodem, S2C, BlueComm, JANUS, ATM	Long-range acoustic underwater communication	External acoustic messaging between AUVs, USVs, or seabed units

This data model and the required services connect via a common connectivity framework, based on the Data Distribution Service (DDS) standard. The scope of UMAA standards is represented graphically in Figure 5-3.

Another US Navy initiative is the Unmanned Maritime Common Control (UMCC), which aims to standardise vehicle planning and control across all unmanned maritime platforms, both on sea surface and underwater.[13]

Control architecture and middleware aspects for AUVs will be covered in the subsequent chapter on Autonomy, Control and Human Interaction.

---

[13]   *Ibid.*

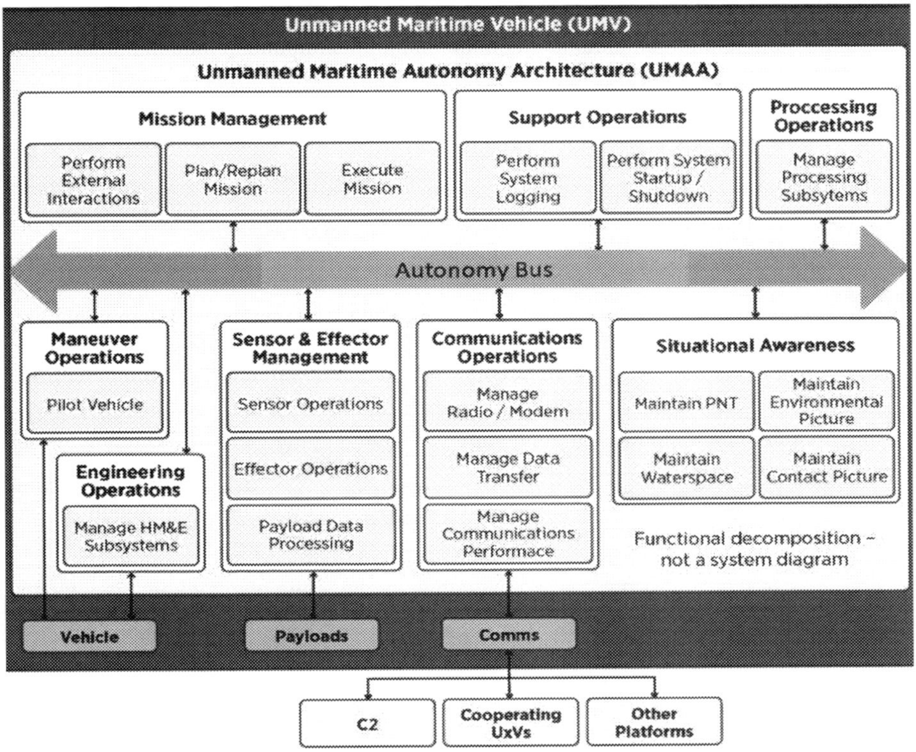

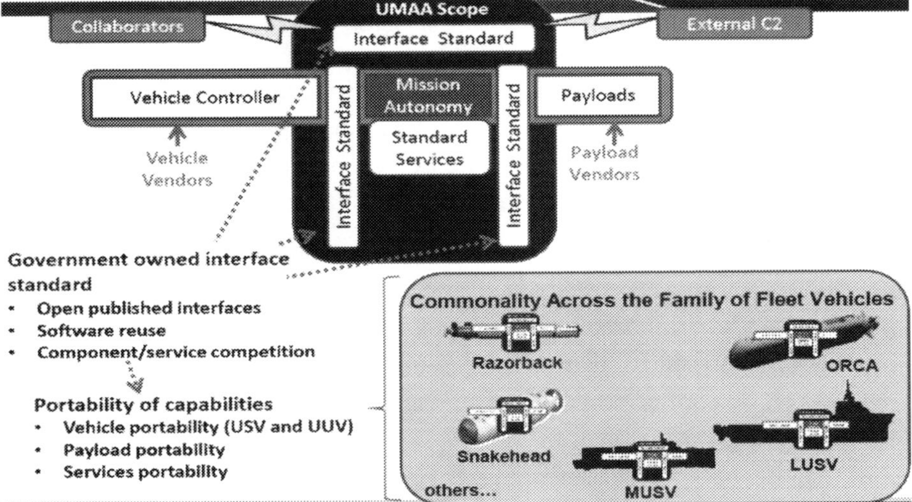

Figure 5-2. Typical top level functional decomposition as per UMAA[14]

Figure 5-3. Scope and objectives of UMAA Standards[15]

---

[14] Program Executive Office, Unmanned and Small Combatants (PEO USC), 2023

[15] *Ibid.*

Software Architecture Standards

In addition to physical design standards, underwater vehicles rely on sophisticated software architectures to manage control, navigation, sensor fusion, and communications. Software architecture is a crucial part of the autonomous operation of AUVs. Examples of standards relevant for design and development of software are mentioned here.

JAUS

JAUS (Joint Architecture for Unmanned Systems) is a specification introduced by the US DoD to enable interoperability between unmanned systems.[16] Developed under SAE AS-4 standards, JAUS is a message-based, platform-independent standard for the command and control of unmanned systems, especially in military/defence AUVs. It defines a set of standard interfaces, messages, and component behaviour to ensure interoperability across unmanned systems in different domains: land, sea and air.

This standard will be discussed further in chapter 9 when we look at software architecture and various middleware including MoOOS-IvP and ROS.

NMEA 2000

One important requirement is to provide for mission-critical data with multiple priorities. This requirement is addressed through proper application of multiple message priority levels of established by product certification. NMEA 2000 has been created by the marine electronics industry as an industry open standard. This is a standard for communication between marine sensors and electronic devices. It defines the physical layer and communication protocol for networking maritime instruments.

The NMEA 2000 network accommodates navigation equipment, electrical power generation and distribution systems, engines and other machinery, piloting and steering systems, fire and other alarms, and controls. Data, commands, and status all share the same cable. Thus, multiple electronic devices can be connected together on a common channel for the purpose of sharing information easily.

---

[16]   Whitsitt & Sprinkle, 2011

Key to NMEA 2000 is an integrated circuit implementation of a network access protocol commonly known as CAN (Controller Area Network). CAN operates in the Carrier-Sense/Multiple Access Mode (CSMA), similar to the alternative protocol of Ethernet. The difference between CAN and the Ethernet approach is seen when two devices simultaneously determine that the bus is not busy and both start to transmit, leading to 'bus contention'. With Ethernet, there is a data collision and both devices stop transmitting and try again later. Valuable time on the bus is lost during the collision and net bandwidth is reduced. CAN handles bus contention in a way that prevents loss of bus bandwidth. The device with the highest priority prevails and continues to transmit data on bit-by-bit basis; the device with the lower priority waits. No data transmission time is lost.[17]

IEEE 12207: Software Life Cycle Processes

IEEE 12207 is the internationally recognized standard for software life cycle processes. It provides a framework for the planning, development, operation, and maintenance of software systems:

- **Systematic Development:** This standard ensures that software is developed in a structured manner, with well-defined phases for requirements analysis, design, coding, testing, and maintenance.

- **Applicability to Underwater Systems:** For underwater vehicles, IEEE 12207 is critical because software failure can lead to catastrophic system failures in harsh marine environments. Adherence to this standard improves reliability and maintainability.[18]

IEEE 14764: Software Maintenance

IEEE 14764 focuses on the maintenance phase of software life cycle processes. It addresses the following aspects:

- **Change Management:** Guidelines for managing software updates, bug fixes, and modifications over the lifetime of the system.

- **Sustainability:** This is particularly important for long-endurance UUVs, which must operate reliably over many years in remote locations.

---

[17] Spitzer, 2009

[18] IEEE 12207: International Standard- Systems and software engineering- Software life cycle processes, 2017

- **Ensuring Safety:** Regular updates and maintenance are crucial to ensure that mission-critical systems continue to operate safely.[19]

Other Relevant IEEE/ISO Standards

Examples of other IEEE standards and ISO/IEC documents impacting the software architecture of underwater vehicles are as follows:

- **IEEE 829:** The Standard for Software Test Documentation, which outlines testing procedures that ensure robust software performance.

- **ISO/IEC 15288:** Just as ISO/IEC 12207 covers systems and software life cycle processes, ISO/IEC 15288 provides frameworks for system life cycle processes. They are widely applied in safety-critical systems and are relevant to the development of UUV control systems.

- **IEEE 1012:** A standard for system and software verification and validation (V&V), ensuring that the final system meets design requirements.

These standards, along with best practices for agile development and real-time system design, are increasingly being adopted by developers of unmanned underwater systems to ensure the integrity and reliability of control software.

Conclusion

Underwater submersibles and UUVs operate in an environment where safety, reliability, and performance are critical. Over the past decades, a robust framework of standards and rules has evolved, driven by classification societies such as ABS, DNV, ClassNK, and LR; software and systems engineering standards from IEEE, ISO, and IEC; and material and mechanical standards from ASME and ASTM. These standards cover everything from the structural design of pressure vessels to the software architectures that control autonomous underwater vehicles and the rugged underwater connectors that ensure reliable communication.

Rapid technological advancements in autonomous systems, sensor integration, and artificial intelligence require further development in standards for software certification, cyber security, and modular interoperability. Enhanced standards and certification processes will help ensure that as

---

[19]   IEEE Standard 14764: Software Engineering - Software Life Cycle Processes - Maintenance, 2017

technologies evolve, vehicles operating beneath the waves remain safe, reliable, and capable of meeting the increasingly inter-connected challenges of scientific exploration, industrial application and defence requirements.

In the next chapter, we move to the energy and propulsion aspects of AUV design and development.

References

(2024). *ABS Requirements for Autonomous and Remote Control Functions.*

(2025). *ABS Rules for Building and Classing Underwater Vehicles, Systems and Hyperbaric Facilities.*

(2019). ASME Boiler and Pressure Vessel Code—Section VIII. ASME.

ASTM F2016-00: Shipbuilding Quality Requirements. (2018). ASTM International.

ClassNK. (2021). *Guidelines for ROVs and AUVs.* Tech. rep., ClassNK.

DNV, G. L. (2015). *Rules for Classification: Underwater Technology—DNVGL-RU-UWT* Pt.3, Ch.2.

DNV, G. L. (2018). *Rules for Classification: Underwater Technology—DNVGL-RU-UWT* Pt.3, Ch.3.

DNV, G. L. (2018). *Rules for Classification: Underwater Technology—DNVGL-RU-UWT* Pt.4, Ch.7.

(2013). *IEC 60529: Degrees of Protection Provided by Enclosures (IP Code).* Tech. rep.

(2017). *IEEE 12207: International Standard- Systems and software engineering- Software life cycle processes.* ISO/IEC/IEEE.

(2017). IEEE Standard 14764: Software Engineering—Software Life Cycle Processes—Maintenance. IEEE.

(2017). *ISO 12944: Corrosion Protection of Steel Structures by Protective Paint Systems.* Tech. rep., ISO.

LR. (2020). *LR Code for Unmanned Marine Systems.* Tech. rep., Lloyds Register.

(2008). *MIL-DTL-38999L: Detail Specification: Connectors.* US Department of Defense.

(2019). *MIL-STD-810H: Test Method Standard: Environmental Engineering Considerations And Laboratory Tests.* Tech. rep., US Department of Defense.

(2018). *Rules for Building and Classing Underwater Vehicles, Systems and Hyperbaric Facilities.* Tech. rep., American Bureau of Shipping (ABS).

Spitzer, S. (2009). NMEA 2000: Past, Present and Future. *RTCM 2009 Annual Conference.* St. Petersburg, Florida.

Unmanned Underwater Vehicles (UUV) Industry Research Report 2023-2030: Opportunities in Deep-Sea Exploration, Subsea Resource Mapping, Maritime Surveillance and Defense Applications. (2024, September 3). Retrieved from: https://www.globenewswire.com/news-release/2024/09/03/2939731/28124/en/Unmanned-Underwater-Vehicles-UUV-Industry-Research-Report-2023-2030

US Department of Defense. (2013). *Unmanned Systems Integrated Roadmap FY2013-2038.* Washington DC: Office of the Under Secretary of Defense (Acquisition Technology and

Logistics). Retrieved from https://ntrl.ntis.gov/NTRL/dashboard/searchResults/titleDetail/ADA592015.xhtml

Whitsitt, S., & Sprinkle, J. (2011). Message modelling for the joint architecture for unmanned systems (JAUS). *18th IEEE International Conference and Workshops on Engineering of Computer-Based Systems*, (pp. 251–259). Las Vegas, NV. doi: 10.1109/ECBS.2011.17

6

Energy and Propulsion

For undertaking their missions in harsh underwater environments without human intervention, energy becomes a fundamental limiting factor. In most AUV systems, the energy available on board constrains mission duration, range, and the quality and quantity of data that can be collected.[1] Therefore, the design, selection, and management of energy sources and power systems for AUVs are of paramount importance.

This chapter discusses key aspects of AUV powering and energy management. It provides an overview of energy sources for AUVs, including batteries, fuel cells, and other alternatives. The typical energy loads encountered during AUV operation (propulsion, hotel loads, payload, communication, etc.) are mentioned.

Propulsion solutions for underwater vehicles are listed, with greater emphasis on thrusters, which are the most prevalent. Steps for selection of thrusters are outlined.

Power generation and distribution systems used onboard and emerging trends in energy harvesting and underwater recharging technologies are covered thereafter, followed by the trade-offs and design challenges that drive the selection of energy systems for underwater vehicles.

---

[1]   Moore, Bohm, & Jensen, 2010

Overview of AUV Energy Systems

Importance of Energy Management

For an AUV, energy is not just a resource; it is the lifeblood that determines operational capabilities. A typical AUV must power several systems, listed below.

- **Propulsion:** Overcoming hydrodynamic drag is energy intensive, and propulsion power largely determines an AUV's range and speed.

- **Hotel loads:** These include onboard computers, sensor systems, communication devices, navigation instruments, and lighting systems. Although individually these loads may be modest, collectively they can contribute a significant share of the overall power consumption.

- **Payload sensors:** Advanced sensor suites (such as imaging systems, sonar arrays, chemical sensors, etc.) have their own power requirements, especially when operating at high resolutions or sampling rates.

Energy management involves balancing these power demands while optimizing the energy storage system so that mission objectives are met, whether they are long-range surveys, deep-sea sampling, or rapid-response operations.

Challenges in Underwater Powering

The major challenges in powering of AUVs are as follows:

- **Limited Onboard Energy:** Due to space, weight, and buoyancy constraints, only a limited amount of energy can be carried on board. This restricts mission duration and range.

- **High Propulsion Demands:** The energy required for propulsion increases with the cube of the speed in many cases. Therefore, any high-speed manoeuvres can rapidly deplete batteries.

- **Harsh Environmental Conditions:** The underwater environment is corrosive, cold (or warm), and under high pressure at depth. Energy storage devices and power electronics must be designed or selected to operate reliably under these conditions.

- **Weight and Volume Constraints:** Sensors, batteries, and power distribution systems all add mass and occupy volume. Maintaining neutral buoyancy and ensuring hydrodynamic efficiency requires careful trade offs.

- **Integration Complexity:** Power distribution must be managed across multiple subsystems with varying voltage and current requirements. In addition, many AUVs must include energy management systems that monitor the state of charge, temperature, and overall health of the energy source in real time.

Energy Sources for AUVs

AUV energy sources have evolved over the decades as battery technologies and alternative power systems have advanced.[2] The two broad categories of energy sources used in AUVs are batteries and alternative systems (such as fuel cells). Other emerging possibilities include hybrid systems and recharging solutions.

Table 6-1 summarises the available solutions for energy storage on AUVs, and they are discussed further.

Table 6-1. Energy Storage Solutions for AUVs[3]

Energy Storage Type	Energy Density (Wh/kg)	Weight Range (Typical for AUV use)	Cycle Life	Operating Temperature Range
Lithium-Ion (Li-ion)	100–250	5–50 kg per module (depending on capacity)	500–1000 cycles	–20°C to +60°C (with proper BMS)
Lithium-Polymer (LiPo)	100–240	100–240 2–30 kg (modular, thin format)	300–800 cycles	–10°C to +50°C
Nickel-Metal Hydride (NiMH)	60–120	10–60 kg	300–500 cycles	0°C to +45°C
Lead-Acid	30–50	20–100 kg (bulky for given capacity)	200–300 cycles	0°C to +40°C
Fuel Cells (PEM)	400–1000 (theore-tical)	Varies widely; typically higher than batteries	5000+ hours (continuous operation possible)	–10°C to +50°C (with proper management)

---

[2] REMUS AUV

[3] Bradley, Feezor, Singh, & Sorrell, 2001; Griffiths, 2003

Batteries

Batteries are the most common energy source for AUVs because of their relative simplicity, high energy density, and maturity of the technology. The major battery chemistries used in AUVs are summarised below.

Lithium-Ion (Li-ion) Batteries

Most modern AUVs (e.g., REMUS 300, Bluefin-21) use Li-ion batteries (Figure 6-1). These can sometimes be enclosed in pressure-tolerant packaging to avoid necessity of locating them inside the pressure-resistant portion of the UUV.[4]

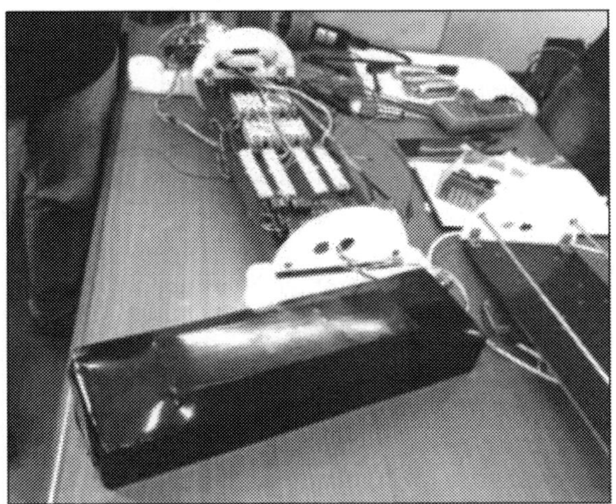

Figure 6-1. Lithium-Ion Battery Pack (Black) being integrated in a small AUV
(Source: NIO Goa)

- **Advantages**
    - o High energy density (typically 100–250 Wh/kg)
    - o Low self-discharge rates
    - o Relatively low weight
    - o Proven performance in many portable electronic and electric vehicle applications

---

[4]   Bluefin Robotics Unmanned Underwater Vehicles

- **Disadvantages**
    - o Sensitivity to temperature and overcharge/over-discharge
    - o Requires sophisticated battery management systems (BMS)
    - o Safety concerns under high stress if not properly protected

Lithium-Polymer (LiPo) Batteries

LiPo batteries are often used in small or micro-AUVs, and for applications where design flexibility and weight savings are critical.

- **Advantages**
    - o Similar energy density to Li-ion but can be manufactured in flexible, lightweight, and thin formats
    - o Can be designed to be pressure tolerant, reducing the need for heavy pressure vessels
- **Disadvantages**
    - o More sensitive to mechanical damage and puncture
    - o Requires careful handling and charging protocols
    - o Lesser cycle life

Nickel-Metal Hydride (NiMH) Batteries

NiMH batteries can be seen in older or budget-constrained designs; they are less common in modern high-end AUVs.

- **Advantages**
    - o Robust and tolerant to overcharge conditions
    - o Lower cost than Li-ion
- **Disadvantages**
    - o Lower energy density (typically 60–120 Wh/kg)
    - o Higher self-discharge rates

Lead-Acid Batteries

Widely used on submarines, lead acid batteries are rarely used in modern AUVs, except where cost is an overriding factor and weight is less critical (e.g., large manned submersibles).[5]

---

[5] Dell & Rand, 2001

- **Advantages**
  - o Inexpensive and proven technology
  - o Robust and tolerant of harsh conditions
- **Disadvantages**
  - o Very low energy density (typically 30–50 Wh/kg)
  - o Heavy and bulky
  - o Requires maintenance and has a limited cycle life

Alternative and Hybrid Energy Systems

While batteries remain the predominant energy source, other technologies are emerging and expected to mature in the years to come. Greater breadth of mission profiles for current and future naval UUVs require longer endurance stealthy propulsion systems that extend the current capability of 10-40 hours to several days or weeks. Current and future anticipated technologies based solely on high energy density batteries may not provide adequate endurance for such missions.

Fuel Cells

Proton Exchange Membrane (PEM) fuel cells are among the most common for marine applications. These are being used on experimental and long-endurance AUVs, especially where extended missions are required and weight constraints can be relaxed. Fuel storage within the vehicle and its refuelling are major challenges.

- **Advantages**
  - o High energy density and longer operational durations
  - o Can be refuelled more rapidly than batteries can be recharged
- **Disadvantages**
  - o Complexity of integration and fuel storage (hydrogen and oxygen)
  - o Higher cost and potentially larger system footprint

Hybrid Systems

Hybrid systems are at the research and development stage, and are intended to combine batteries with fuel cells or super-capacitors. They are being tried out on advanced AUV designs targeting both high peak performance and extended endurance. According to DARPA, advanced UUV power supply technology based on hybrid fuel cells would be capable of operating up to four hours on maximum power and operating under ideal conditions for 1,680 hours.[6]

In a hybrid fuel cell system, a hydrogen fuel cell is supported by a traditional battery. The hydrogen fuel cell would use polymer electrolyte membrane (PEM) technology, where a combination of cathodes and anodes separate the protons and electrons of the hydrogen molecules and the PEM forces the electrons to travel through a circuit to generate power. The smaller, traditional battery would offer supplemental power when required by the size of the payload. Meanwhile the hydrogen fuel cell would provide power at a stable rate for the steady load and recharge the small battery during low power intervals.[7]

- **Advantages**
  - o Batteries can provide high peak power for manoeuvres, while fuel cells supply long-term energy
  - o Supercapacitors can handle transient loads and regenerative braking
- **Disadvantages:**
  - o Increased system complexity
  - o Integration challenges in managing multiple energy sources simultaneously

Other Emerging Alternatives

- **Underwater Wireless Power Transfer:** Techniques such as magnetic induction or acoustic energy transfer are being explored for recharging AUVs while deployed. These systems are in the early research phase but hold promise for reducing downtime by allowing in situ recharging.[8]
- **Solar and Thermal Energy:** While solar energy is impractical for deep-sea missions, AUVs that surface periodically can harness solar panels to extend endurance. Additionally, ocean thermal energy conversion (OTEC) is being explored to capture the temperature gradient between surface and deep waters.

---

[6]   Owens, 2022

[7]   Navy SBIR, 2015

[8]   Wang, et al., 2024

- **Microbial Fuel Cells:** Utilize the metabolic processes of bacteria to generate electricity from organic matter in seawater. Although currently limited by low power output, they offer an intriguing possibility for long-term, low-power sensor applications.

Typical Energy Loads

AUV energy consumption can be divided into the following major categories:

- **Propulsion Load:** The power required to overcome hydrodynamic drag and move the vehicle.
- **Hotel Load:** Energy consumed by onboard systems such as sensors, communications, navigation, and computing.
- **Payload Load:** Power drawn by mission-specific equipment (e.g., imaging systems, chemical sensors).

Propulsion Power

Propulsion is typically the most energy-intensive part of an AUV's operation. The power required is influenced by:

- **Hydrodynamic Drag:** Generally increases with the cube of the velocity.
- **Efficiency of the Propulsion System:** This includes the motor, gear train (if any), and propeller. Propeller efficiency, transmission losses, and motor control all affect overall power consumption.
- **Operating Speed:** A balance must be struck between speed and energy consumption. Most AUV missions are designed to operate at low speeds (e.g., 1–3 knots) to maximize endurance.

Typical propulsion loads for mid-range AUVs vary widely. High-speed manoeuvres may require power levels in tens of kilowatts (e.g., 20–50 kW) during short bursts. The majority of operations are at a lower cruising speed, requiring power in the range of 1–5 kW, depending on vehicle size and hydrodynamic design.[9]

Hotel and Payload Loads

Hotel loads include the continuous energy demand of electronic systems that maintain control, communication, and sensor operation. Typical values include:

---

[9]　Griffiths, 2003; Roberts & Sutton, 2013

- **Onboard Computing and Control:** 5–30 W.
- **Sensor Suites:**
  - o Environmental sensors (CTD, dissolved oxygen, etc.): 0.1–1 W per sensor when duty-cycled.
  - o Active sensors (sonar, high-definition cameras): May require 5–20 W during active operation.
- **Communication Systems:** Acoustic modems may draw 1–5 W during transmission; high-data-rate systems (if used) can require more.
- **Lighting Systems:** For imaging applications, underwater lights might draw 5–10 W.

The overall hotel load in many AUV designs is typically in the range of 50–100 W (when averaged over the mission), although peak loads during sensor operation or data transmission can be higher.[10]

Table 6-2 summarises typical energy loads due to various AUV systems.

Table 6-2. Typical Energy Loads on AUVs

Subsystem	Power Requirement	Duty Cycle	Remarks
Propulsion (Cruising Mode)	1–5 kW	Continuous during transit	Dominant power consumer; varies with speed and drag
Propulsion (High-Speed Burst)	20–50 kW (for short durations)	Brief bursts during manoeuvres	Peak power demands during acceleration or evasion
Onboard Computing & Control	5–30 W	Continuous (can be throttled down)	Includes navigation, control algorithms, and monitoring
Environmental Sensors (CTD, DO, pH)	0.1–1 W per sensor	Intermittent duty (e.g., 10–30% active)	Often duty-cycled to conserve energy
Acoustic Sensors (Active Sonar)	5–20 W	Intermittent during active scanning	Power varies with ping rate and intensity
Optical Imaging (HD Camera)	1–5 W (low-light mode)	Continuous or triggered based on need	Higher for high-resolution video; may be duty-cycled
Communication (Acoustic Modem)	1–5 W	Variable (higher during data transmission)	Often includes standby and active modes
Additional Payload (e.g., Chemical Sensors)	0.5–2 W each	Typically low duty cycle	Specific to mission requirements

---

[10] AUV ROV Sensors; Roberts & Sutton, 2013

AUV Propulsion Technologies

Propulsion solutions for AUVs can be broadly classified as follows:

- Conventional thruster propulsion
- Buoyancy-driven propulsion
- Wave-generated motion
- Biomimetic propulsion

Comparison of Propulsion Approaches

Each propulsion type has advantages and trade-offs, enumerated below.

Thruster-Based Propulsion

Electric thrusters, typically using brushless DC (BLDC) motors coupled directly to a propeller, are the workhorse of many AUV designs. Many modern thrusters feature fully flooded designs that allow water to cool and lubricate internal components, eliminating the need for complex sealing. Popular examples include the Blue Robotics T200 and T500 units.[11]

Advantages

- **Rapid Response and Precision:** Direct electric thrusters enable quick acceleration and fine manoeuvring.
- **Modular and Scalable:** Thruster arrays can be arranged to provide full six-degree-of-freedom control.
- **Mature Technology:** Extensive field data and commercial availability make this a reliable option.

Limitations

- **Energy Intensive:** Continuous operation leads to higher power consumption.
- **Potential Noise and Vibration:** Despite improvements, some noise and cavitation remain, which may affect sensitive sensors.

Details of thruster selection are covered in the next sub-section.

Buoyancy-Driven Gliders

Underwater gliders use variable buoyancy systems to achieve a sawtooth dive profile. Hydrodynamic wings convert the vertical motion into forward glide.

---

[11] Blue Robotics Thrusters, Actuators, and Lights

This method is exceptionally energy efficient, enabling missions that span weeks or even months, as seen for platforms like Slocum and Seaglider (Figure 6-2).[12]

Figure 6-2. 'Seaglider' Underwater Glider of APL,
University of Washington[13]

Advantages

- **Exceptional Endurance:** Minimal power usage allows for missions lasting weeks to months.
- **Low Noise:** Lack of continuously running motors results in quiet operation.

Limitations

- **Lower Speeds:** Typical velocities are much lower compared to thruster-driven AUVs.
- **Limited Agility:** Not suitable for rapid manoeuvring tasks.

Wave-Generated Motion

Some unmanned surface vehicles (USVs) exploit ocean wave energy. The Wave Glider (Figure 6-3), for example, utilizes the relative motion between a

---

12 Ray, Singh, & Seshadri, 2011; Graver, 2005
13 Seaglider Autonomous Underwater Vehicle; Ray, Singh, & Seshadri, 2011

surface float and a submerged glider to produce thrust without fuel consumption.[14] This method offers extended endurance and sustainable operation.

Advantages

- **Renewable Energy:** Harvesting wave energy eliminates fuel requirements.

- **Extended Mission Duration:** Capable of long-term operation with minimal intervention.

Limitations

- **Environmental Dependence:** Performance is contingent on consistent wave activity.

- **Reduced Control Precision:** Generally slower and less manoeuvrable than thruster-driven systems.

Figure 6-3. Wave Glider developed by Liquid Robotics[15]

---

[14] WaveGlider: Persistent Ocean Monitoring

[15] *Ibid.*

Biomimetic Propulsion

Biomimetic systems replicate the efficient swimming mechanisms of marine animals—such as the caudal fin motions of fish or jet propulsion in squids—to potentially offer superior efficiency and manoeuvrability with lower noise. Although many biomimetic designs remain in the experimental phase, they are promising for applications that demand high agility and stealth.[16] Computational fluid dynamics (CFD) plays a crucial role in optimizing these complex designs.

Advantages

- **High Efficiency and Low Noise:** Mimicking biological locomotion can yield quieter and more efficient systems.
- **Adaptive Performance:** Potential for superior manoeuvrability in complex flow conditions.

Limitations

- **Design Complexity:** Replicating natural motion involves intricate mechanical and control challenges.
- **Early Development Stage:** Many systems remain in the prototype phase, with commercialization yet to be achieved.

Thruster-Based Propulsion Overview

Since thrusters are the most widely used propulsion method, this section looks at details of their design, performance, and control.[17]

Design Principles

Conventional thrusters employ sensor-less BLDC motors with fully flooded housings that allow water circulation for cooling and lubrication. This design not only enhances pressure tolerance but also reduces maintenance requirements by eliminating traditional seals. While most systems use a direct mechanical coupling between the motor and propeller, recent developments include magnetic or reconfigurable couplings that permit thrust vectoring without rotating the entire assembly.

---

[16] Bandyopadhyay, 2005; Li, et al., 2023
[17] Wadoo & Kachroo, 2011; Hasan, 2017

Performance Parameters

Key performance metrics for thrusters include:

- **Power Range:** Small units may operate in the 50–500 W range, while larger thrusters can require several kilowatts.
- **Voltage and Current:** Typical operating voltages range from 7 V to 20 V (with some high-power units operating at higher voltages). For example, the T200 thruster draws approximately 17–32 A, whereas the T500 draws around 43.5 A at 24 V.
- **Thrust Output:** Expressed in kgf or newtons, thrust levels vary from a few kgf (for small AUVs) to over 16 kgf in high-power models.
- **Coupling and Noise:** Direct coupling is standard; however, advanced magnetic couplings can offer dynamic thrust vectoring. Fully flooded designs help minimize cavitation and mechanical noise.

Control Systems

Electronic speed controllers (ESCs) use pulse-width modulation (PWM) to manage power delivery, ensuring efficient motor operation across a range of loads. Modern systems often integrate ESCs with onboard sensors to adjust parameters in real time for optimal performance and to safeguard against faults.

Commercial Examples

Table 6-3 summarises specifications for a few representative commercially available thruster models (example shown in Figure 6-4) for AUV/ROV applications.

Table 6-3. Typical commercial AUV thruster specifications[18,19]

Thruster Model	Operating Voltage	Power Range	Current Rating	Thrust Output	Coupling Mechanism	Key Features
Blue Robotics T200	7–20 V	~205–645 W	~17 A (12 V) – 32 A (20 V)	~3.7–6.7 kgf (8.2–14.8 lbf)	Direct, fully flooded	Compact, affordable BLDC design; widely used in platforms like BlueROV2
Blue Robotics T500	16–24 V	~1 kW+	~43.5 A at 24 V	~16.1 kgf (35.5 lbf)	Direct, fully flooded	High-power with ~3× T200 thrust, rugged; pressure tolerant
Tecnadyne AUV Thrusters	Typically 12–48 V	50 W – 11 kW	Varies (10s to 100s of A)	From 10s to 100s of kgf	Direct; with redundant seals	Engineered for work-class AUVs; robust in harsh conditions

---

18   Blue Robotics: https://bluerobotics.com/product-category/thrusters/
19   Tecnadyne: https://tecnadyne.com/thrusters

Figure 6-4. Blue Robotics Thruster T500[18]

Thruster Sizing and Selection

Choosing the correct thruster involves careful consideration of vehicle characteristics, mission requirements, and environmental conditions. This section outlines a methodology for thruster sizing and selection.

Sizing Criteria

The process begins with establishing performance requirements:

- **Mission Profile:** Define the desired speed, manoeuvrability, and endurance. High-speed missions require higher thrust; long-duration surveys favour energy-efficient units.

- **Vehicle Mass and Hydrodynamics:** Determine the total mass, drag characteristics, and resistance forces acting on the AUV. This involves calculating the required net thrust to overcome drag at the intended operating speed.

- **Redundancy and Manoeuvring:** In multi-thruster configurations, consider both the aggregate thrust and the ability to vector thrust for six-degrees-of-freedom control.

Engineers typically use performance charts provided by manufacturers— comparing thrust output versus current draw and voltage—to estimate the required power. CFD simulations and experimental validation can further refine these estimates by modelling flow interactions around the vehicle hull and thruster inlets.

Selection Process

A systematic selection process involves the typical steps shown in the flowchart in Figure 6-5. This approach ensures that the propulsion system is neither over-designed (resulting in unnecessary energy consumption and weight) nor underpowered for the vehicle's mission profile.[20]

As an example, consider an AUV intended for a survey mission with a cruise speed of 1.5 m/s:

- **Step 1:** Use CFD or empirical drag models to determine that the vehicle requires 15 kgf of net thrust.

- **Step 2:** Evaluate thruster models (e.g., Blue Robotics T200 vs. T500) and find that a T200 unit (with thrust output up to 6.7 kgf) arranged in a multi-thruster configuration (e.g., four units) can achieve the necessary thrust with added redundancy.

> **Estimate Drag:**
> Use CFD or empirical formulas to determine the drag force at the desired cruising speed

> **Estimating Thrust Needs:**
> To the estimated drag at the maximum desired speed, add a margin to account for manoeuvring and unexpected conditions

> **Reviewing Manufacturer Specifications:**
> Compare candidate thrusters and choose candidates that meet or exceed the required thrust while operating within the desired voltage and current ranges

> **Simulation and Prototyping:**
> Incorporate candidate thrusters into CFD models of the AUV. Estimate/simulate the selected thruster's performance within the AUV's hull configuration and perform controlled experiments to verify the sizing assumptions

> **Integration and Testing:**
> Validate the selection through bench tests and sea trials to confirm that the thruster meets both dynamic and endurance requirements

Figure 6-5. Suggested flowchart for thruster selection (Author)

---

[20] Rutherford, 2008

- **Step 3:** Simulate the integration of these thrusters on the vehicle hull to verify that the added drag and flow interference remain within acceptable limits.

- **Step 4:** Perform sea trials to fine-tune control parameters and confirm energy efficiency.

Selection Guidelines

When selecting thrusters, apart from meeting the peak speed (thrust) requirement, the following aspects should also be considered:

- **Match to Vehicle Dynamics:** The thruster's thrust curve should align with the vehicle's acceleration and deceleration needs. For instance, an AUV requiring rapid course corrections may benefit from thrusters with a low response time and high efficiency at partial loads.

- **Efficiency Over Operating Range:** Evaluate efficiency curves to ensure that the thruster operates near its peak efficiency during most of the mission.

- **Physical Integration:** Size and weight constraints are critical. Thrusters must fit within the vehicle's design envelope without compromising buoyancy or balance.

- **Environmental Considerations:** Choose thrusters with suitable pressure ratings and corrosion-resistant materials, especially for deepwater or harsh marine environments.

- **Control and Redundancy:** For mission-critical applications, consider systems with built-in redundancy and advanced control capabilities (e.g., magnetic-coupling units that offer thrust vectoring).

An integrated approach combining CFD analysis, bench testing, and performance modelling is recommended to optimize thruster selection.

Power Generation and Distribution

Power is distributed across the AUV via a centralized DC bus, with DC-DC converters used to supply the correct voltage to each subsystem. Key considerations for power distribution architecture include the following:

- **Isolation and Redundancy:** To protect sensitive systems from voltage spikes or faults in other parts of the system.

- **Efficiency:** High-efficiency converters are essential to minimize losses, especially given the limited energy resources onboard.
- **Scalability:** Modular power distribution systems allow new subsystems to be added with minimal redesign.

Power Generation

For AUVs, the most common method of power generation is through onboard batteries. In recent designs, the focus has shifted towards increasing energy density and reducing the weight and volume of battery packs. Some of the options are mentioned below.

- **Battery Pack Configuration:** Battery packs are arranged in series or parallel (or both) to achieve the desired voltage and capacity. Modern systems use high-discharge Lithium-ion or Lithium-polymer (LiPo) cells to meet peak power demands.
- **Fuel Cells:** Fuel cells offer an alternative with a high energy density and the advantage of fast 'refuelling' (or fuel replacement) compared to recharging batteries. These systems are being considered for long-duration missions.
- **Hybrid Systems:** Hybrid solutions use a combination of batteries and super capacitors. The batteries supply continuous energy, while the super capacitors handle high-power bursts (such as for propulsion surges), thereby protecting the battery from stress.

Power Distribution

Once the energy is generated or stored, it must be distributed to the various subsystems in the AUV:

- **DC Power Bus:** Most AUVs use a DC bus architecture where the battery output is regulated and distributed through DC-DC converters to the different modules (propulsion, sensors, computing, etc.).
- **Power Management Electronics:** These systems monitor battery state-of-charge, temperature, and health. They also control load prioritization and implement low-power modes during standby.
- **Redundancy:** In safety-critical applications (especially in military AUVs), power distribution systems are designed with redundancy. Multiple power channels and backup batteries ensure that a failure in one channel does not jeopardize the entire mission.

Table 6-4 summarises typical AUV power generation and distribution systems.

Table 6-4. Overview of AUV Power Generation and Distribution

System Type	Description	Typical Voltage Range	Distribution Methods	Advantages	Disadvantages
Direct Battery Power	Supply DC power directly to subsystems	12–48 V typical	DC bus with DC-DC converters	Simplicity; mature technology	Limited by battery capacity; weight constraints
Fuel Cell-Based Systems	Generate electricity continuously, often in hybrid with batteries	24–48 V (depending on configuration)	Integrated power electronics, DC bus with conversion	Extended endurance; fast refuelling possibilities	Complex integration; higher cost
Hybrid Systems (Battery + Super-capacitor)	Combines high-energy batteries with supercapacitors for peak load management	12–48 V typical	Power management system with energy buffering	Smoothens transient load; extends battery life	Increased system complexity
Wireless Power Transfer	Underwater charging via magnetic induction or acoustic coupling	Variable	Dedicated charging coils/ docking interface	Potential for in situ recharging; reduces downtime	Currently low efficiency; short range; still in R&D

Energy Management Strategies

Effective energy management is vital for maximizing the operational efficiency and endurance of an AUV. Strategies include both hardware-based approaches and intelligent software control.

Duty Cycling and Load Scheduling

Many AUV subsystems do not require continuous operation. Energy usage can be optimised by the AUV mission computer through various measures.

- **Sensor Duty Cycling:** Environmental sensors can be switched on intermittently, based on predefined sampling intervals or triggered by environmental changes.

- **Variable Propulsion Modes:** An AUV might operate in low-speed, energy-saving 'survey' mode and then switch to high-speed bursts only when necessary.

- **Standby Modes:** Non-critical systems can enter a standby state to conserve energy during long transit periods.

Onboard Energy Monitoring

Modern AUVs employ sophisticated battery management systems (BMS) that continuously monitor voltage, current, temperature, and state-of-charge. They provide real-time data to the mission control software, allowing adaptive adjustments to power allocation. They can also alert operators to potential issues and automatically adjust system performance to conserve energy.

Battery management systems (BMS) provide real-time feedback to the AUV's control system, enabling:

- **State-of-Charge Estimation:** Accurate estimation of remaining battery capacity.

- **Thermal Management:** Active regulation of battery temperature through insulation, passive cooling, or even active liquid cooling.

- **Fault Detection:** Early identification of battery or power distribution faults, allowing the AUV to take corrective action.

Power Optimization Algorithms

AUVs are beginning to use adaptive algorithms that adjust the power distribution based on real-time measurements and mission phases. Energy management algorithms play a critical role in reducing overall power consumption. They should dynamically allocate power to subsystems based on mission phase and load demands.[21] Such adaptive power management should include the following features:

- **Predictive Energy Management:** Onboard software can predict energy consumption based on mission profiles (historical data) and real-time sensor inputs and adjust operating parameters (speed, sensor activation, etc.) to optimise battery life.

- **Optimise Mission Profiles:** Modify speed, path, and operational modes dynamically to minimize energy use while still achieving mission objectives.

---

[21] *Ibid.*

- **Adaptive Control:** Incorporate machine learning to improve efficiency over time, learning from previous missions to better schedule sensor activation and propulsion bursts.
- **Dynamic Load Adjustment:** Non-critical systems can be throttled or shut down during periods of high energy demand.

Thermal Management

Battery performance and longevity are highly dependent on temperature. Thermal management strategies include:

- **Insulation:** Battery compartments are often insulated to maintain optimal operating temperatures.
- **Active Cooling:** In some designs, small pumps or fans circulate coolant around the batteries.
- **Thermal Coupling:** Using waste heat from propulsion systems or onboard electronics to warm batteries in cold environments.

Futuristic Developments

In addition to improvements in battery and fuel cell technologies, researchers are exploring innovative methods for extending AUV endurance through energy harvesting and underwater recharging.

Energy Harvesting

Energy harvesting technologies aim to capture ambient energy from the underwater environment, supplementing onboard energy reserves. Some promising approaches are mentioned below.

Wave and Current Energy Harvesting

Devices (often based on piezoelectric or electromagnetic induction principles) convert the kinetic energy of water motion into electrical energy.[22]

- **Applications:** Useful for AUVs operating in areas with strong currents or wave action. Such systems might not supply high power but can extend mission duration by trickle-charging onboard batteries.
- **Challenges:** Harvesting systems must be robust, lightweight, and efficient at converting low-frequency, low-amplitude energy fluctuations.

---

[22] WaveGlider: Persistent Ocean Monitoring

**Figure 6-6. The 3-metre long 'Wave Glider' by Liquid Robotics
uses solar and wave energy**[23]

Solar Power

Solar panels integrated on AUVs that surface periodically can harvest solar energy (Figure 6-6). Although sunlight attenuates rapidly underwater, solar cells can be effective during surface intervals.[24]

- **Applications:** Gliders and other AUVs that spend a significant amount of time near the surface can benefit from solar recharging.
- **Challenges:** Panels must be robust, waterproof, and optimized for variable light conditions.

Microbial Fuel Cells

This concept aims to utilize the metabolic activity of marine bacteria to generate electricity from organic matter in seawater.[25]

- **Applications:** Particularly promising for long-term environmental monitoring systems that require only low power.
- **Challenges:** Current output levels are very low, and scaling up remains a technical challenge. However, advances in biotechnology may improve efficiency.

---

[23] *Ibid.*

[24] *Ibid.*

[25] Massaglia, 2021

Underwater Recharging

Underwater recharging technologies aim to allow AUVs to replenish their energy supplies without the need to return to a base or surface vessel.

- **Wireless Power Transfer (WPT):** Magnetic Induction through coils could be used to transfer power via a magnetic field (Figure 6-7). Although efficient over very short distances, research is ongoing to increase the effective range.[26]

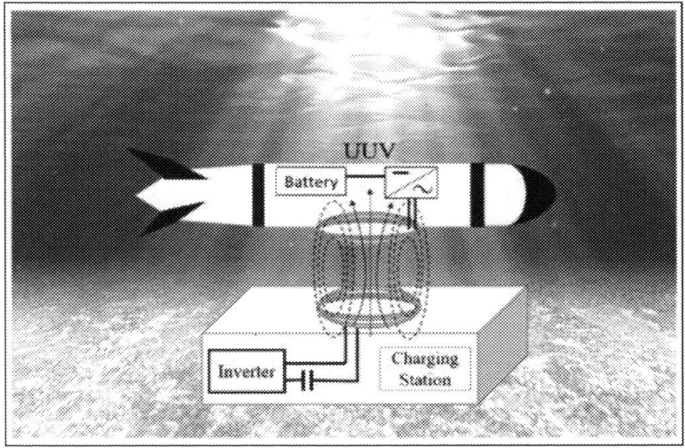

Figure 6-7. Concept of underwater wireless charging system for UUVs[27]

- **Mechanical Docking:** AUVs could be designed to autonomously dock with an underwater charging station. These stations can provide power via direct electrical contacts or through inductive coupling.[28]

- **Automated Battery Swapping:** Some futuristic designs propose modular battery packs that can be automatically exchanged by an onboard mechanism during a docking procedure.

- **Standardisation of recharging modules:** Standardised interfaces for energy storage and recharging modules would enable interoperability and easier upgrades.

- **Hybrid Systems:** Combining traditional batteries with energy-harvesting modules and in situ recharging infrastructure could offer a pathway to continuous or near-continuous operation.

---

[26]   Wang, et al., 2024

[27]   Mohsan, Khan, Mazinani, Alsharif, & Cho, 2022

[28]   Liu, Yu, He, & Soares, 2024

Integration with Internet of Underwater Things

The future of underwater energy management may also involve networked sensor and vehicle systems, referred to as Internet of Underwater Things (IoUT) or Subsea Internet of Things (SIoT).[29] Such underwater networks would not be able to use electromagnetic waves, and without physical connections, would need to rely on acoustic communication, whose limitations would govern the viability of the concept.

- **Edge Computing and Data Fusion:** Onboard processors can optimize energy usage by processing sensor data locally and transmitting only essential information, reducing communication energy costs.

- **Cloud-Connected Recharging:** Integration of AUVs with a subsea IoT may allow for centralized energy management, remote monitoring, and even coordinated recharging strategies across a fleet of vehicles.

Conclusion

Energy is the most critical resource for Autonomous Underwater Vehicles, and its management defines the operational capabilities and mission success of these platforms. The evolution of AUV energy systems from traditional battery packs to fuel cell and hybrid configurations reflects the ongoing challenge of providing sufficient power in a harsh underwater environment. The quest for more efficient, reliable, lightweight, and high-density energy storage systems in the Electric Vehicles market would have positive spin-offs for AUVs.

Advanced on-board energy management strategies, such as duty cycling, adaptive power allocation, and robust battery monitoring systems would ensure that AUV energy is used as efficiently as possible.

AUV propulsion is a critical determinant of mission success. Conventional thrusters, with their proven responsiveness and modularity, continue to be the preferred solution for applications requiring agility and speed. However, buoyancy-driven gliders, wave-powered systems, and emerging biomimetic approaches offer advantages in energy efficiency and stealth.

---

[29]   Nkenyereye, Nkenyereye, & Ndibanje, 2024

Looking forward, advances in CFD, AI-driven optimization, and novel materials are likely to further refine thruster-based systems and accelerate the integration of hybrid propulsion solutions. Such innovations could enhance energy efficiency, reduce environmental impact, and expand the operational capabilities of AUVs, enabling new applications in deep-sea exploration and long-duration surveillance.

Energy harvesting from the ocean and recharging from underwater docking stations are some of the exciting technologies that hold great promise for extending reach and endurance of underwater systems. However, for underwater nodes to be able to function as a swarm, a network, or an IoT, the reliability and range of underwater communication becomes the limiting factor. In the next chapter, we examine the challenges and current maturity of underwater navigation and communication technologies.

References

AUV ROV Sensors. (n.d.). Retrieved May 2025 from Sea-Bird Scientific: https://www.seabird.com/auv-rov-sensors/family

Bandyopadhyay, P. (2005, January). Trends in Biorobotic Autonomous Undersea Vehicles. *IEEE Journal of Oceanic Engineering, 30*, 109-139. doi:10.1109/joe.2005.843748

Blue Robotics Thrusters, Actuators, and Lights. (n.d.). Retrieved May 2025 from Blue Robotics: https://bluerobotics.com/product-category/thrusters/

Bluefin Robotics Unmanned Underwater Vehicles. (n.d.). Retrieved May 2025 from General Dynamics Mission Systems: https://gdmissionsystems.com/underwater-vehicles/bluefin-robotics

Bradley, A.M., Feezor, M.D., Singh, H., & Sorrell, F.Y. (2001). Power systems for autonomous underwater vehicles. *IEEE Journal of Oceanic Engineering, 26*(4), 526–538.

Dell, R.M., & Rand, D.A. (2001). *Understanding Batteries*. Royal Society of Chemistry. doi:10.1039/9781847551065

Graver, J. (2005). Underwater glider: dynamics, control and design. *Underwater glider: dynamics, control and design*. Princeton University.

Griffiths, G. (ed.). (2003). *Technology and Applications of Autonomous Underwater Vehicles*. Taylor & Francis. doi:10.1201/9780203023249

Hasan, H. (2017). *Underwater Robotics: Science, Design and Fabrication*. Malaysia: UTM Press.

Li, G., Liu, G., Leng, D., Fang, X., Li, G., & Wang, W. (2023). Underwater Undulating Propulsion Biomimetic Robots: A Review. *Biomimetics, 8*(3), 318. doi:10.3390/biomimetics8030318

Liu, J., Yu, F., He, B., & Soares, C.G. (2024, April). A review of underwater docking and charging technology for autonomous vehicles. *Ocean Engineering, 297*(1). Retrieved May 2025 from https://www.sciencedirect.com/science/ article/abs/pii/S0029801824004918

Massaglia, G. (2021). Integration of Portable Sedimentary Microbial Fuel Cells in Autonomous Underwater Vehicles. *Energies, 14*. Retrieved May 2025 from https://www.researchgate.net/

publication/353525521_ Integration_of_Portable_Sedimentary_ Microbial_Fuel_Cells_in_ Autonomous_Underwater_Vehicles

Mohsan, S.A., Khan, M.A., Mazinani, A., Alsharif, M.H., & Cho, H.-S. (2022, September). Enabling Underwater Wireless Power Transfer towards Sixth Generation (6G) Wireless Networks: Opportunities, Recent Advances, and Technical Challenges. *Journal of Marine Science and Engineering*, 10(9). doi:https://doi.org/10.3390/jmse10091282

Moore, S., Bohm, H., & Jensen, V. (2010). *Underwater Robotics: Science, Design and Fabrication*. Monterey, CA: Marine Advanced Technology Education Center.

Nkenyereye, L., Nkenyereye, L., & Ndibanje, B. (2024). Internet of Underwater Things: A Survey on Simulation Tools and 5G-Based Underwater Networks. *Electronics (Special Issue: Artificial Intelligence Empowered Internet of Things)*, 13(3). doi:https://doi.org/10.3390/ electronics13030474

Qiu, T., Zhao, Z., Zhang, T., Chen, C., & Chen, C. (2020). Underwater Internet of Things in Smart Ocean: System Architecture and Open Issues. *IEEE Transactions on Industrial Informatics*, 16, 4297–4307.

Ray, A., Singh, S., & Seshadri, V. (2011). Underwater Gliders: Force Multipliers for Naval Applications. *Warship 2011: Naval Submarines and UUVs*. Bath, UK: Royal Institution of Naval Architects. doi:10.3940/rina.ws.2011.03

REMUS AUV. (n.d.). Retrieved May 2025 from Woods Hole Oceanographic Institution: https://www.whoi.edu/what-we-do/explore/underwater-vehicles/auvs/remus/

Roberts, G. N., & Sutton, R. (eds.). (2013). *Further Advances in Unmanned Marine Vehicles*. The Institution of Engineering and Technology. doi:10.1049/PBCE002E

Rutherford, K. T. (2008). *Autonomous Underwater Vehicle Design Considering Energy Source Selection and Hydrodynamics*. University of Southampton.

Seaglider Autonomous Underwater Vehicle. (n.d.). Retrieved May 2025 from Applied Physics Laboratory (APL), University of Washington: https://apl.uw.edu/project/ project.php?id=seaglider

Tecnadyne Thrusters: Product information. (n.d.). Retrieved May 2025 from Tecnadyne: https:/ /tecnadyne.com/thrusters

Wadoo, S., & Kachroo, P. (2011). *Autonomous Underwater Vehicles: Modelling, Control, Design, and Simulation*. CRC Press, Taylor & Francis Group.

Wang, D., Zhang, J., Cui, S., Bie, Z., Chen, F., & Zhu, C. (2024, January). The state-of-the-arts of underwater wireless power transfer: A comprehensive review and new perspectives. *Renewable and Sustainable Energy Reviews*, 189(A). doi:https://doi.org/10.1016/ j.rser.2023.113910

WaveGlider: Persistent Ocean Monitoring. (n.d.). Retrieved May 2025 from Liquid Robotics: https://www.liquid-robotics.com/wave-glider/how-it-works/

7

Underwater Navigation and Communication

Navigating beneath the surface of the water poses a unique set of challenges. Radio-frequency signals attenuate rapidly in water. Therefore, unlike land or aerial vehicles, underwater systems cannot rely on the Global Positioning System (GPS) or on radio communication. Instead, modern AUVs need to use a combination of inertial sensors, acoustic ranging systems, Doppler velocity logs (DVLs), and, occasionally, geophysical cues to estimate their position. The many challenges underwater include sensor drift, multipath effects in acoustics, and the highly dynamic nature of the underwater environment.

In the context of UUVs, 'navigation' denotes estimating the true position, velocity, and attitude of the vehicle, accounting for uncertainty and disturbances. The first part of this chapter presents an overview of underwater navigation methods used in modern AUVs. The benefits of sensor fusion and processing improvements for reducing drift and increasing navigation accuracy are discussed. The accuracy figures reported from various field trials and experiments are summarised.

The next part covers the challenges and developments in underwater communication, which is predominantly acoustic. Over decades, engineers and researchers have developed specialised techniques to transmit, receive and process acoustic data underwater. These technologies are critical for scientific oceanography, environmental monitoring, and industrial applications including offshore energy and infrastructure inspection. Naturally, they are essential for naval applications as well, such as mine countermeasures, anti-submarine warfare, and surveillance.

We examine the primary communication modalities used underwater—acoustic, optical, and electromagnetic—and evaluate their performance in terms of range, bandwidth, and reliability. We also review emerging hybrid solutions and discuss how developments in materials, signal processing, and artificial intelligence (AI) are advancing the limits of the field.

Navigation Techniques and Sensors

Continuous improvements in sensor technologies and sophisticated sensor fusion algorithms have significantly enhanced the accuracy and robustness of underwater navigation. However, it still remains a challenge compared to the accuracy and ease of navigation on land or in the air.

A summary of localisation (i.e. AUV navigation) methods, sensors and approaches is shown in Figure 7-1.[1]

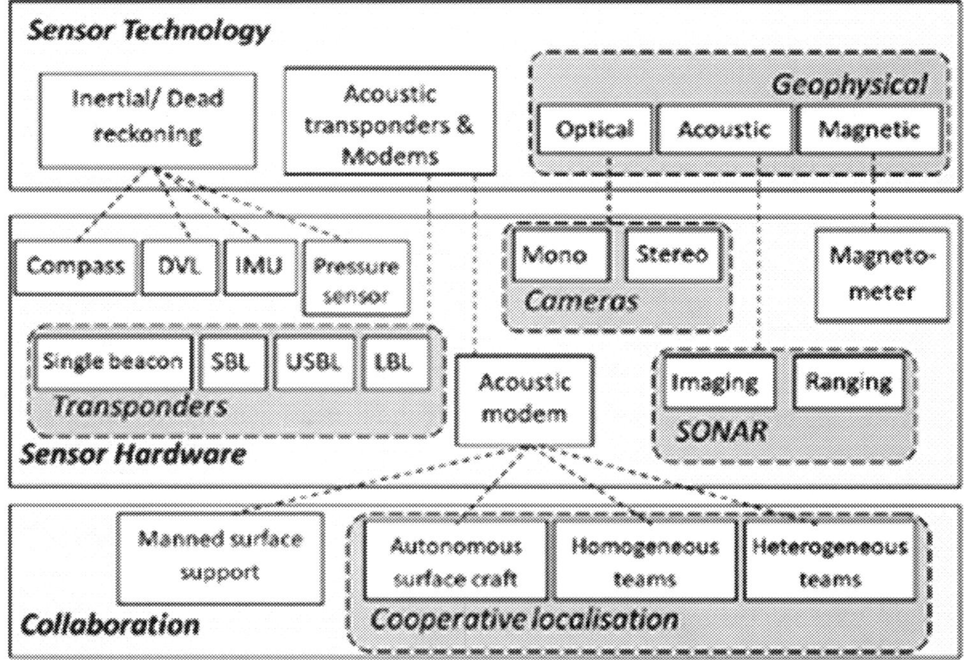

Figure 7-1. Localisation and navigation technologies for AUVs (Paull, 2014)

---

[1]   https://people.csail.mit.edu/lpaull/publications/Paull_JOE_2013.pdf

Recent field trials and experimental campaigns have demonstrated error drifts as low as 0.1–0.2 per cent of the distance travelled. Accuracy of even 0.02 per cent of distance travelled has been reported in a multi-week mission over a range of 1,200 nautical miles.[2] A fusion of an inertial navigation system (INS) with a DVL and acoustic updates has demonstrated drift corrections that yield final position errors in the low tens of metres over missions spanning several kilometres.[3] Processing techniques such as the Extended Kalman Filters (EKF), Unscented Kalman Filters (UKF), and graph-based SLAM (Simultaneous Localisation and Mapping) have played a crucial role in minimising cumulative errors in navigation.

We start with an overview of the major techniques and instruments employed in underwater navigation.

Inertial Navigation Systems (INS) and Dead-Reckoning

Inertial navigation systems use accelerometers and gyroscopes (often integrated into an inertial measurement unit, or IMU) to compute the vehicle's position by integrating the measured acceleration and angular rates. In practice, INS provides continuous navigation information without external references.

However, due to sensor noise, bias, and the integration process, errors (or drift) accumulate over time. In typical AUV missions without external correction, the INS drift may reach tens or even hundreds of metres over several hours. When operating solely on an INS (i.e., dead-reckoning), errors accumulate with time. Without external correction, field trials have reported drifts of the order of several per cent of the distance travelled. For example, a long-duration mission spanning 10 km may accumulate drift errors of the order of 100–200 m if relying on INS alone.[4]

Therefore, INS-only navigation is generally acceptable only for short-duration missions (e.g., in confined laboratory tanks or short coastal surveys).

Doppler Velocity Log (DVL)

DVLs measure the vehicle's velocity relative to the seafloor (or water column) by transmitting acoustic pulses and measuring the Doppler shift of the returned signal. When integrated with an INS, DVL data can substantially reduce

---

[2]   Kongsberg Discovery, 2024

[3]   Paull, Saeedi, Seto, & Li, 2014

[4]   Kinsey, Eustice, & Whitcomb, 2006

drift by providing velocity updates. The quality of the DVL measurements depends on factors such as the altitude above the seafloor, bottom characteristics, and acoustic signal quality. DVLs are available with depth ratings from 300 to 6,000 metres, with bottom track from over 0.1 to 200 m range.

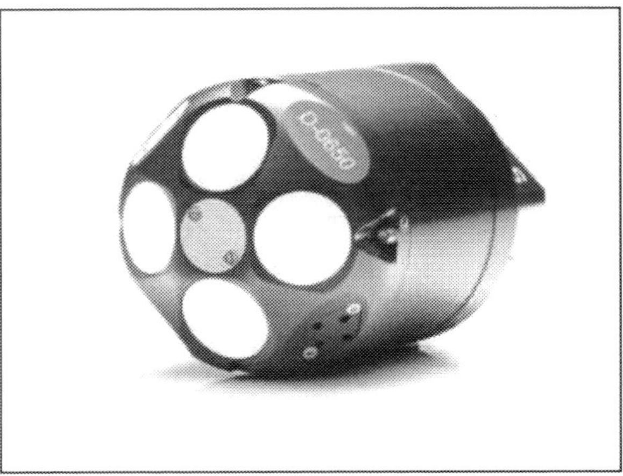

Figure 7-2. Doppler Velocity Log
(Source: Nortek[5])

The integration of DVL data with an INS is the most common approach to reducing drift. Field experiments report that when the DVL is 'bottom-locked' (i.e., when the vehicle is sufficiently close to the seafloor), the combined INS/DVL solution can achieve drift rates as low as 0.1–0.2 per cent of the distance travelled. For instance, a combined INS/DVL system may achieve a navigation error of around 10–20 m over a 10-km mission.[6] In one untethered under-ice mission (approximately 11 km), the majority of the navigation updates had corrections below 15 m.[7]

Kalman Filter-Based Processing
In order to fuse the data from the INS and the aiding sensors, some form of filtering must be implemented. This is typically accomplished using an error-state Kalman filter (KF). A wide range of position aiding tools can be integrated, as well as additional velocity aiding sources.[8]

---

[5]   DVL1000

[6]   Randeni, Schneider, Bhatt, Víquez, & Schmidt, 2022

[7]   Bhatt, Viquez, & Schmidt, 2022

[8]   Paull, Saeedi, Seto, & Li, 2014

A simplified schematic diagram showing the integration of DVL data is shown in Figure 7-3. The KF input is the difference between the output from the appropriate aiding sensors and the INS. The output from the KF includes estimates of the slowly varying systematic errors of the navigation sensors, as well as sea current when using DVL water-track data.

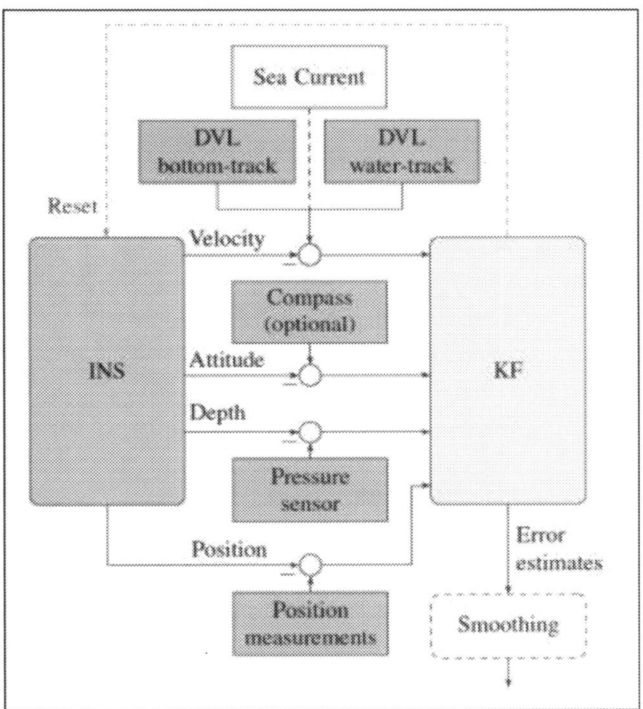

Figure 7-3. Simplified structure of AUV INS with error state KF [9]

Extended Kalman Filters (EKF)

EKF is one of the most widely used algorithms for sensor fusion in underwater navigation. By linearizing the nonlinear models of vehicle dynamics around the current estimate, the EKF can recursively fuse INS, DVL, and acoustic measurements to produce an improved estimate of position and velocity. Field trials have shown that EKF-based INS/DVL systems can reduce drift rates to around 0.1–0.2 per cent of the distance travelled.

---

[9] Hegrenæs, Ramstad, Pedersen, & Velasco, 2016

Unscented Kalman Filters (UKF)

The UKF avoids linearization by propagating a set of sample points (sigma points) through the nonlinear model. In many experiments, UKF has demonstrated a modest improvement over EKF—for example, reducing errors by an additional 10–15 per cent in tests comparing the two methods. In one set of trials, UKF-based fusion reduced the drift error by roughly 0.1 m compared to EKF over short-duration manoeuvres.

Acoustic Positioning Systems

Acoustic positioning systems—such as Long Baseline (LBL), Ultra-Short Baseline (USBL), and Short Baseline (SBL) systems—rely on acoustic transponders and receivers to determine the vehicle's position by measuring the time-of-flight of sound waves.[10] These systems provide absolute position fixes that can be used to correct INS drift. LBL systems use a network of transponders placed on the seafloor, whereas USBL/SBL systems use surface vessels or moored buoys. Reported accuracies vary with configuration; typical absolute positioning errors of 5–15 m are common in many field trials.

Ultra-Short Baseline (USBL)

The USBL includes two modes—the transponder mode, and the responder mode. In the transponder mode, an interrogation signal is sent from the surface transducer array to the seabed transponder.[11] After this interrogation signal is received, an acknowledge signal is transmitted from the transponder. The time delay between the transmission of the interrogation signal and the reception of the acknowledge signal is used for calculating the range.

In the responder mode, a pair of high-precision clocks is used to trigger the responder on the submersible and USBL surface processing unit on the mother ship synchronously. The time delay between the synchronous pulse triggering moment and the reception of the acknowledge signal is used for calculating the slant range. The sound speed profile is needed for calculating the range from the time delay. Thus, an accurate sound speed profile guarantees accurate positioning. In addition to the range, the azimuth is needed for positioning and can be computed by analyzing the phase of the received

---

[10]   Kongsberg Maritime, 2023
[11]   L3Harris, 2022

signals. Combining the GPS and USBL, the exact location of the AUV (or submersible) can be obtained.

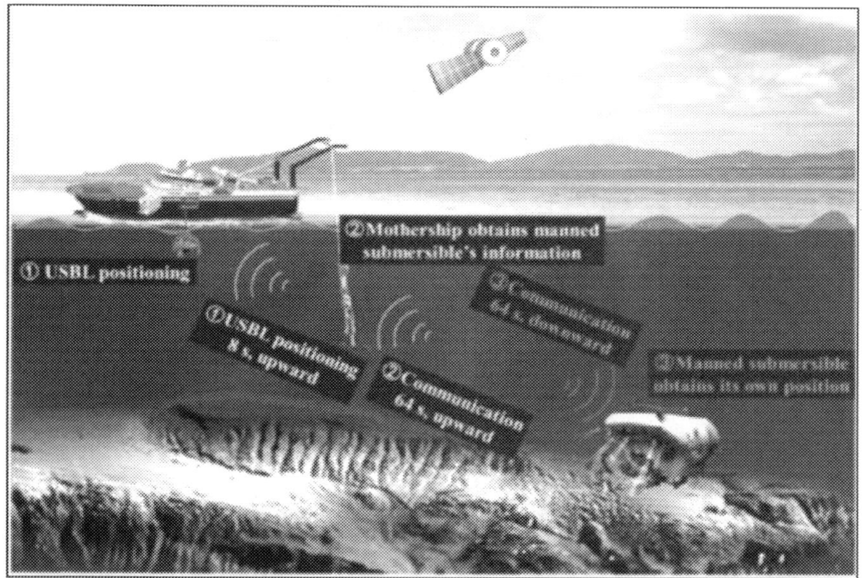

Figure 7-4. Steps for a Submersible to obtain its position using USBL[12]

A submersible (as illustrated in Figure 7-4) can obtain its position through the following three steps using USBL:

- The responder beacon on the submersible transmits a positioning pulse at regular intervals (for example, 8 s). The position of the submersible can be obtained after the USBL on the mother ship receives the pulse.

- The submersible sends its status information to the mother ship at longer time intervals (for example, 64 s) through digital underwater acoustic communication.

- After the mother ship receives this status information, it sends submersible position information calculated by the USBL to the submersible through feedback communication.

---

[12]   Zhang, Liu, & Liu, 2018

Long Baseline (LBL)

The LBL system consists of two parts: a transducer and processing unit installed in the manned submersible, and a set of beacons at the bottom of the ocean. The LBL acoustic positioning converts the travel times to ranges utilizing three or more widely spaced (approximately 4 km) stationary beacons for calculating the location of a moving vehicle in two or three dimensions. The beacons are typically stationary (fixed baseline) and moored to the seafloor with tethers. Alternatively, they can be held at fixed relative locations on a moving platform, such as a ship (moving baseline). At least three beacons (known positions) are required to form a specific geometry. The beacon locations must be determined during an initial offline survey.[13]

Figure 7-5. Position measurement by LBL using underwater beacons[14]

A schematic diagram of LBL positioning is shown in Figure 7-5. The LBL transducer sends an interrogation signal to each beacon and receives acknowledge signals from the beacons, which are rendered uniquely identifiable by generally assigning unique coded pulses or frequencies to each beacon. The time delay can then be measured, and the ranges between the LBL transducer and each beacon can be calculated.

---

[13] *Ibid.*
[14] *Ibid.*

Geophysical Navigation

Geophysical navigation utilizes natural environmental gradients—such as bathymetry, geomagnetic anomalies, or gravity fields—to estimate the vehicle's position by matching sensor observations to pre-existing maps.[15] While geophysical methods can potentially provide long-term drift correction, their accuracy depends on the resolution and quality of the environmental map, as well as the sensitivity of the sensors used.

Using the spatial and temporal distribution characteristics of the seafloor topography and the underwater magnetic field as navigation beacons, an optimized positioning algorithm framework can be constructed. This could use EKF to couple the topographic elevation information and magnetic field strength information, with the seafloor topography and the marine geomagnetic map. This can be used to determine the relative position of the AUV with respect to the marine geophysical information chart, and to correct the accumulated positioning error of the dead reckoning (Figure 7-6).[16]

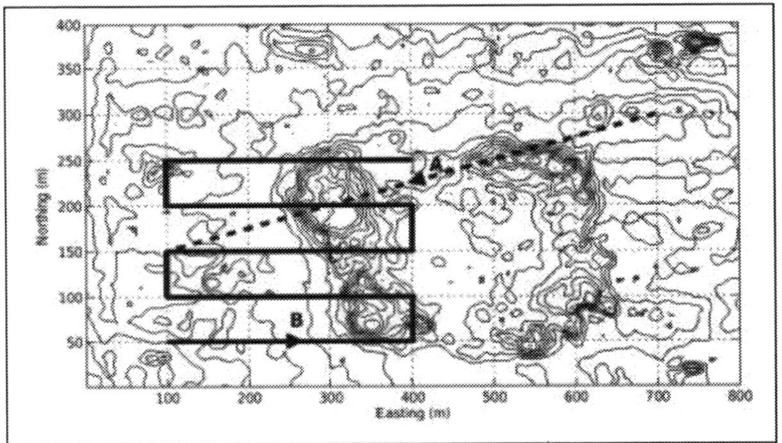

Figure 7-6. Localisation of trajectories based on bathymetric contour map[17]

When high-resolution bathymetric or magnetic maps are available, geophysical navigation methods can correct for INS drift, with reported accuracies varying from a few metres to tens of metres, depending on the feature resolution.[18]

---

[15]   Meduna, Petillot, & Lane, 2008

[16]   Yu, Zhang, Chen, Huang, & Lan, 2023

[17]   *Ibid.*

[18]   Teixeira & Pascoal, 2007

Optical and Visual-Based Navigation

In clear water conditions, visual navigation methods using cameras or lasers can provide high-resolution, short-range measurements of the environment.[19] Techniques such as visual odometry and simultaneous localization and mapping (SLAM) have been successfully applied in controlled environments (e.g., in lakes or clear ocean regions). However, in turbid or deep-water environments, optical methods are limited by light attenuation and scattering.

Despite the success of SLAM in terrestrial environments, its adaptation to underwater domains presents significant challenges. These include low visibility due to light absorption and scattering, sensor noise, dynamic lighting conditions, and distortions caused by the water column. Unlike in air, where visual and LiDAR-based SLAM systems are widely used, underwater SLAM relies on a combination of sonar, acoustic signals, inertial sensors, and vision-based approaches, each with inherent limitations. These constraints degrade the accuracy of feature extraction, loop closure detection, and trajectory estimation, which are critical components of SLAM pipelines.

In clear water conditions, visual SLAM systems have reported relative localization errors of the order of 20–60 cm in controlled environments; however, these figures deteriorate rapidly in turbid or dynamic settings.

Particle Filters and Graph-Based SLAM

Methods such as particle filters and graph-based SLAM provide robustness in highly nonlinear, non-Gaussian environments, enabling loop closure and long-term drift correction. Although computationally demanding, they are useful in extended missions where error accumulation is a major concern.

Particle filters have been used in underwater navigation to handle non-Gaussian noise and non-linear dynamics. Although computationally more intensive, particle filters can better manage uncertain and intermittent sensor updates (e.g., when DVL measurements are partially lost). In simulation studies, particle filter-based fusion methods have reduced cumulative errors significantly compared to traditional EKF approaches, especially in scenarios with dynamic acoustic conditions.

Graph-Based SLAM formulates the SLAM problem as a graph optimisation task. These approaches have shown good performance in reducing

---

[19]   Caccia, 2006

drift by optimising the entire trajectory through loop closure detection.[20] When integrated with INS/DVL systems for long-term navigation, graph-based methods help correct accumulated drift by re-aligning the trajectory with previously mapped features, yielding overall position errors of a few metres.[21]

Traditional SLAM methods are primarily model-based probabilistic algorithms using mathematical tools like Bayesian filters (EKF, particle filters), graph optimisation, and geometric computer vision (Table 7-1).

Table 7-1. Examples of traditional SLAM methods in underwater environments[22]

Reference	*Sensor*	*Front-End*	*Back-End*	*Focus*	*Findings*	*Limitations*
Bonin-Font et al. (2015)	Stereo Cameras	—	Graph-SLAM & EKF-SLAM	Localization, Mapping	Graph-SLAM out performs EKF-SLAM	Limited by imaging conditions
Demim et al. (2022)	Sonar	Hough Transformation	ASVSF SLAM	Localization, Mapping	Better accuracy than EKF-SLAM	Needs further validation
Rahmati et al. (2019)	Generic	SURF	SLAM with Adaptive Sampling	Navigation, Mapping	Efficient data collection in water bodies	Limited by tether dependence
Zhang et al. (2022)	Optical Cameras	ORB Feature Detection	ORB-SLAM2	Localization	Effective for underwater robot localization	Requires distortion correction
Carrasco et al. (2015)	Stereo Cameras	—	Graph-SLAM	Navigation, Localization, Control	Stereo vision improves localization	Computationally intensive
Palomeras et al. (2019)	Multi-beam Sonar	ICP Algorithm	Active SLAM	Localization, Mapping	Maintains vehicle uncertainty bounded	Limited by environmental variability

Deep Learning-based SLAM

Traditional SLAM can be considered part of the broader AI field but is generally not categorized as ML unless combined with data-driven learning techniques. Recent advances in deep learning (DL) have introduced powerful

---

[20] Eustice, Singh, Leonard, Walter, & Ballard, 2005

[21] Yuan, Martínez, Pons, & Eckert, 2017

[22] Heshmat, et al., 2025

solutions to attempt improvement of accuracy and robustness of underwater navigation systems in real world conditions.

DL techniques enhance underwater SLAM by improving feature extraction, image de-noising, distortion correction, and sensor fusion.[23] Convolutional neural networks (CNNs) and transformer-based architectures have demonstrated remarkable improvements. Multi-modal approaches integrating vision, sonar, and inertial data have proven effective in compensating for sensor deficiencies, enabling more resilient localization and mapping systems.

Integration of UWSN

In addition to enhancing SLAM, an emerging research direction is the integration of underwater wireless sensor networks (UWSNs). UWSNs provide a distributed sensing framework where multiple AUVs and sensor nodes collaboratively share localization and mapping data via acoustic communication.[24] By incorporating UWSN-based SLAM, navigation accuracy and robustness can be improved, especially in large-scale, multi-agent underwater operations.

Sensor Fusion and Processing

One of the main avenues for achieving high accuracy in underwater navigation is the effective integration of multiple sensors using advanced fusion techniques. The inherent drift of INS-only systems makes it imperative to integrate external sensors that provide absolute or relative fixes. Sensor fusion combines the short-term accuracy of INS with the long-term stability of external references (e.g., DVL, acoustic positioning, and geophysical measurements). These methods combine the strengths of different sensors to mitigate individual weaknesses, particularly drift in inertial sensors and limited ranges of acoustic systems.

Sensor Fusion Examples

By sharing sensor data among vehicles and employing decentralized fusion algorithms, the collective system is less prone to localized sensor failures or drift. The combination of INS and DVL has been discussed earlier. We now look at other examples of sensor data fusion for navigation.

---

23 *Ibid.*
24 Chaudhary, et al., 2023

INS/Acoustic Positioning Fusion

Acoustic positioning systems such as LBL, USBL, or SBL are used to provide absolute position fixes that can be fused with INS data to correct drift.[25] Fusion with acoustic positioning systems is particularly beneficial in scenarios where the vehicle operates away from the seafloor (and therefore may lose DVL lock) or in dynamic environments where environmental drift (such as ice drift) must be compensated. The acoustic updates help to constrain the error growth when the vehicle is operating outside the range of the DVL.[26]

Field trials in under-ice environments have been reported in the Beaufort Sea (ICEX20) over an 11-km mission. By using a combined INS/DVL/acoustic update system (HydroMAN-ICNN framework), 59 per cent of navigation correction updates were found to be less than 15 m, and 77 per cent of updates were below 20 m error.[27]

In other long-range missions of up to 60 km by Autosub3 beneath the Antarctic ice shelf, the incorporation of USBL updates reduced cumulative errors to within 20 metres, although such performance is highly dependent on environmental conditions.[28]

Multi-Sensor Fusion in Collaborative Projects

European projects such as CoCoRo[29] have demonstrated that in multi-agent systems, distributed sensor fusion not only allows for cooperative mission planning and obstacle avoidance, but also provides robust long-term drift correction. The combination of INS, DVL, acoustic ranging, and environmental sensors (e.g., CTD sensors) yielded overall position accuracies typically within 15–20 m in multi-hour missions of several kilometres.[30]

Improvements in Processing

Advances in processing hardware and algorithms have further contributed to reducing navigation errors:

- **Real-Time Processing:** With the advent of high-performance embedded processors and GPUs, sensor fusion algorithms that were

---

25   Tiano & Ferri, 2018

26   Asai, Kojima, Asakawa, & Iso, 2000

27   Randeni, Schneider, Bhatt, Víquez, & Schmidt, 2022

28   Bhatt, Viquez, & Schmidt, 2022; McPhail, 2009

29   COCORO, 2015

30   Randeni, Schneider, Bhatt, Víquez, & Schmidt, 2022

once too computationally demanding can now be run in real time on AUVs.[31]

- **Adaptive Filtering:** Modern fusion algorithms often include adaptive elements that adjust the filter's parameters in response to changes in sensor quality or environmental conditions. Adaptive filtering algorithms can adjust noise covariance matrices in real time based on sensor performance and environmental conditions. This dynamic tuning helps maintain high accuracy even when some sensors degrade temporarily (e.g., when a DVL loses bottom-lock momentarily).

- **Outlier Rejection:** Robust statistical techniques and machine learning methods have been integrated to detect and reject erroneous sensor measurements before they corrupt the fusion process. Outlier rejection algorithms—such as those based on statistical methods or machine learning (e.g., using Voronoi diagrams for USBL data)—ensure that erroneous measurements do not corrupt the sensor fusion process. By discarding spurious data, the fusion filter can maintain a tighter bound on drift and error accumulation.

- **Multi-Sensor Redundancy:** Combining measurements from multiple sensors provides redundancy that reduces the overall uncertainty and drift. Redundancy in sensor systems (e.g., having both upward- and downward-looking DVLs, or multiple acoustic transducers) further contributes to improved reliability and accuracy. In several field trials, this redundancy has allowed systems to 'reset' accumulated errors periodically, effectively capping the maximum drift over long missions.

Summary of Navigation Systems

As mentioned at the beginning of the chapter, in the context of Guidance, Navigation, and Control (GNC) for UUVs, navigation is generally understood as the process of determining the vehicle's state vector, including its position, velocity, and attitude (orientation).[32] This encompasses what is often called 'localisation' or 'positioning', meaning the determination of the vehicle's location and orientation in its environment.

Future underwater navigation systems should be able to achieve higher levels of accuracy and reliability by developments such as adaptive filtering,

---

[31] Bachmayer et al., 2008
[32] Fenucci, 2024; Fossen, 2002.

integration of emerging sensors, computational efficiency and cooperative navigation.[33] These should enable more complex and extended missions in diverse underwater environments.

The systems available to an AUV for determining its position, as discussed in this chapter, are summarised and compared qualitatively in Table 7-2.

Table 7-2. Characteristics of navigation / positioning systems for UUVs[34]

Positioning System	Advantages	Disadvantages
Inertial Navigation System (INS)	• No need for external signals • Can operate in GPS-denied environments	• Drift over time leading to decreased accuracy • Requires periodic calibration
Long Baseline (LBL)	• High positional accuracy • Effective in deep water • Suitable for complex environments	• Requires installation of multiple transponders • Limited mobility during set up
Short Baseline (SBL)	• Less complex than LBL • Suitable for shallower waters • Easier and quicker to deploy	• Lower accuracy compared to LBL • Limited range and precision
Ultra-Short Baseline (USBL)	• Real-time tracking-Requires only a single transducer • Good for dynamic environments	• Less accurate at greater distances • Performance can be affected by surface conditions
Doppler Velocity Log (DVL)	• Provides velocity relative to the seafloor or water column • Enhances positioning accuracy if combined with other systems	• Accuracy affected by water and seafloor characteristics • Not a standalone positioning system
Dead Reckoning	• Simple and cost-effective • Useful for estimating position in the absence of other data	• Accuracy degrades over time • Dependent on accurate initial conditions
Acoustic Modem Systems	• Enables communication between UUVs and surface vessels • Can assist in positioning and data transfer	• Limited bandwidth-Communication can be affected by water conditions and distance
LIDAR (Light Detection & Ranging)	• Provides high-resolution 3D data • Effective for shallow-water mapping	• Limited depth penetration • Requires clear water conditions

Underwater Communication

Unlike terrestrial or satellite communication, underwater channels are characterized by high signal attenuation, variable sound speed, and complex

---

[33] Bahr, Leonard, & Fallon, 2009; Leonard & Bahr, 2016.

[34] Alexandris, Papageorgas, & Piromalis, 2024

multipath propagation.[35] Most of the research and development in ocean acoustics, and acoustic communication, was driven by military applications of sonar and improvements to detect submarines and ships.[36]

The rise of autonomous underwater vehicles (AUVs) has further driven innovation in underwater communication.[37] Autonomous systems require robust, low-latency, and energy-efficient communication links to share sensor data, receive mission updates, and coordinate with other platforms. We start by reviewing the communication requirements or use cases for various applications.

Current underwater communication technologies can broadly be categorized into three primary modes: acoustic, optical, and electromagnetic (EM). Each has unique advantages and limitations, which dictate its suitability for different applications.[38] For each of these three modes, the subsequent sections will summarise the characteristics (frequency range, modulation techniques), performance metrics (range, bandwidth and latency) and maturity in various applications (naval, scientific and industrial).

Requirements in various Applications

Underwater communication technologies are critical to a wide range of scientific, industrial and naval (military) applications.[39] Certain key application areas and examples of how these technologies are deployed are mentioned here.

Scientific Applications

- **Oceanographic Research:** AUVs and sensor networks used for mapping the seafloor, studying ocean currents, and monitoring water quality rely heavily on underwater communication for real-time data collection and control.[40] For example, networks of sensor nodes deployed in the Pacific Ocean transmit data on temperature, salinity, and biological activity using low-power acoustic modems.

---

[35]   Busby, 1976

[36]   Moore, Bohm, & Jensen, 2010

[37]   Aoki, 2008

[38]   Kilfoyle & Baggeroer, 2000

[39]   Geyer, 1977; Choi & Yuh, 2016.

[40]   Billon-Coat, 2018; Wynn, Huvenne, Bas, Murton, et al, 2014.

- **Environmental Monitoring:** Monitoring coral reef health, tracking pollutant dispersion, and assessing climate change impacts require reliable communication between autonomous platforms and shore-based laboratories.[41]

- **Deep-Sea Exploration:** High-bandwidth optical systems are required in deep-sea exploration to transmit high-resolution images and video data from the seafloor, allowing researchers to remotely study undersea ecosystems.

Industrial Applications

- **Offshore Energy:** In the oil and gas industry, underwater communication systems support the inspection of pipelines, wellheads, and subsea installations.[42] High-bandwidth optical links are particularly useful for real-time video inspection and diagnostics.

- **Infrastructure Monitoring:** Autonomous underwater sensor networks are deployed to monitor the integrity of underwater cables, bridges, dams and port facilities. Data is transmitted to maintenance teams for proactive repair planning.

- **Resource Exploration:** UUVs used in seabed mining or mineral exploration rely on acoustic communication to transmit high-resolution sonar and geophysical data over moderate distances.

Naval Applications

- **Covert Communication:** Submarines and unmanned underwater vehicles use acoustic communication for covert data transmission. Secure, low-bandwidth channels are critical for maintaining stealth.

- **Network-Centric Warfare:** Modern naval operations increasingly involve distributed unmanned systems that share data via integrated communication networks. This enables coordinated surveillance, mine countermeasures, and strike operations.[43]

- **Manned-Unmanned Teaming:** Autonomous systems deployed from naval vessels use hybrid communication systems to maintain constant

---

41 Tijjani & Chemori, 2021; Roberts & Sutton, 2013.

42 Finn & Scheding, 2010

43 RAND Corporation, 2019

connectivity with command centres, facilitating dynamic mission updates and coordinated responses.[44]

Acoustic Communication

Acoustic communication is the most widely used method underwater due to the relatively low absorption of sound in water. Sound travels in water at approximately 1500 m/s, though this speed varies with temperature, salinity, and depth. Sound waves can propagate long distances, albeit with limitations imposed by multipath effects, ambient noise, and Doppler shifts.[45] Acoustic communication relies on the transmission of sound waves, which can be modulated to carry digital data.

Acoustic Modem

The acoustic modem is the equipment used for underwater wireless communication between a UUV and its mother ship, or between a UUV and other UUVs/ nodes. It works by encoding and sending digital data (e.g., telemetry, commands, or sensor readings) as acoustic signals. A receiver decodes the signals into usable information. It is used for communicating with submerged vehicles, for remote sensor data retrieval, and for command and control in subsea missions. There is a fundamental trade-off—lower frequency corresponds to a longer range, but lower data rate (e.g., 8–15 kHz, ~1–2 kbps). Higher frequency provides shorter range, with higher data rate (e.g., 35–50 kHz, ~20–50 kbps).

Performance is affected by turbulence, noise, thermoclines, and multiple acoustic paths. Directional transducers can improve range and reduce interference. Size of the equipment can vary, with an example of a relatively large one being EvoLogics S2C-R,[46] weighing 2 to 9 kg and size about 110 mm (diameter) x 178 mm. A summary of typical acoustic modem equipment characteristics is given in Table 7-3.

New acoustic modems have been developed that offer higher bandwidths (up to tens of kilobits per second) over moderate ranges (up to a few kilometres).[47] These devices benefit from improved modulation techniques such as OFDM and advanced error correction algorithms.

---

[44]   National Research Council, Committee on Autonomous Vehicles in Support of Naval Operations; Naval Studies Board; Division on Engineering and Physical Sciences, 2005

[45]   Goetz, et al., 2012

[46]   https://www.evologics.com/product/s2c-r-18-34-1

[47]   Hasan, 2017

Table 7-3. Communication performance of typical acoustic modems

Make/Model	Power (Tx/Rx/Idle)	Data Rate	Range
Sonardyne Modem/ 6	—	≤ 9/ kbps	≤ 5/ km
WHOI Micromodem 2	idle ~0.5/ W	80 to 5,400/ bps	~1/ km+
EvoLogics S2C R	2.5/ mW to 65/ W (Tx varies)	13.9 to 31.2/ kbps	1 to 3.5/ km
Water Linked M16	Rx <0.3/ W, Tx <1.5/ W	— (packets)	~1/ km
LinkQuest UWM	Tx 2–8/ W	up to 38.4/ kbps	~1 to 5/ km
Benthos DAT	—	~0.14 to 15/ kbps	2 to 6/ km (extensible)

Acoustic System Characteristics

- **Frequency Range:** Acoustic communication systems typically operate in frequencies ranging from a few hundred hertz to tens of kilohertz. Lower frequencies enable long-range communication (up to tens or even hundreds of kilometres) but provide lower data rates, whereas higher frequencies support higher bandwidth but are limited in range.

- **Modulation Techniques:** Common modulation techniques include Frequency Shift Keying (FSK), Phase Shift Keying (PSK), and more advanced methods such as Orthogonal Frequency Division Multiplexing (OFDM) to maximize data throughput and reliability.

- **Low-Power Acoustic Networks:** Research into low-power acoustic communication has enabled the development of sensor networks for environmental monitoring. These networks can operate for extended periods on battery power, transmitting data intermittently.[48]

- **Adaptive Acoustic Systems:** Advances in signal processing have allowed acoustic systems to adapt to changing environmental conditions, dynamically adjusting frequency, power, and modulation schemes to optimize performance.[49]

Performance Metrics

- **Range:** Low-frequency acoustic communication can achieve ranges from 10 to over 100 km in optimal conditions. However, in practical applications, ranges are often limited to a few kilometres in noisy or shallow environments.

---

[48] Randeni, Schneider, Bhatt, Víquez, & Schmidt, 2022

[49] Kim, Lee, & Myung, 2020

- **Bandwidth:** Typical acoustic communication channels provide bandwidths from a few hundred hertz to several kilohertz, which translates to data rates in the range of a few kilobits per second to tens of kilobits per second. Field tests of modern acoustic modems have demonstrated ranges of up to 10 km in favourable conditions, with effective bandwidths ranging from 1 to 10 kbits per second. Advanced modulation schemes have pushed these limits further, with some experimental systems achieving 20 kbits per second over moderate ranges.

- **Latency:** Acoustic signals suffer from significant latency due to the slow speed of sound in water. This can be of the order of hundreds of milliseconds to seconds, depending on the distance. This is acceptable for many autonomous and monitoring applications but limits real-time control in high-speed operations.

Maturity of Acoustic Communication

- **Naval:** Long-range communication for submerged vessels, covert messaging, and coordinated manoeuvres.

- **Scientific:** Data retrieval from sensor networks, AUV telemetry, and real-time control in oceanographic research.

- **Industrial:** Monitoring and control of subsea installations, remote inspections, and environmental surveys.

Optical Communication

Optical communication underwater uses light (typically lasers) to transmit data. This method can offer very high data rates over short distances. Light propagates in water with significant attenuation due to absorption and scattering. The performance is highly dependent on water clarity; it is most effective in clear, deep water.[50]

Characteristics

- **Wavelengths:** Blue and green wavelengths are preferred because they experience less attenuation in seawater compared to other wavelengths.

---

[50]   Partan, 2007

- **Modulation:** High-speed modulation techniques enable data rates that can reach megabits per second, albeit over ranges typically limited to tens of metres.

- **Optical Wireless Communication (OWC):** OWC systems are being developed for short-range high-bandwidth communication between underwater vehicles and between vehicles and surface stations. Prototypes have demonstrated in controlled conditions.[51]

- **Alignment and Beam Steering Technologies:** Innovations in beam steering and automatic alignment systems could help maintain optical links even in the dynamic underwater environment.

Performance Metrics

- **Range:** Optical communication systems typically achieve ranges of 10–100 metres in clear ocean water. In turbid or coastal waters, effective range may be reduced significantly, often to less than 10 m.

- **Bandwidth:** Optical channels can provide bandwidths from several megahertz up to several gigahertz, supporting high data rates. Optical underwater communication systems have achieved data rates of up to several megabits per second in clear water over distances of 10–50 m.[52]

- **Latency:** Because optical signals travel at the speed of light, latency is extremely low (a few microseconds); however, overall system latency may be affected by processing and alignment issues.

Maturity of Optical Communication

Underwater optical communication for UUVs is primarily at the experimental or prototype stage. In future, this technology could have potential use as follows:

- **Scientific:** High-speed data transfer from AUVs for real-time video, imaging, and sensor data in short-range operations.

- **Industrial:** Inspection and control of underwater infrastructure where high-resolution imagery is required.

- **Naval:** Short-range secure communication between underwater vehicles and between a vehicle and a surface station.

---

[51] Antonelli, 2018

[52] Rice & Frater, 2011

Electromagnetic (EM) Communication

Electromagnetic communication underwater is challenging due to the high attenuation of radio waves in seawater, but it finds niche applications, particularly in shallow or near-surface environments. EM waves in water are quickly absorbed, with the penetration depth inversely related to frequency. Lower frequencies can penetrate further but offer very limited bandwidth.

Characteristics

- **Frequency Range:** Extremely low frequency (ELF) and very low frequency (VLF) signals are used for long-range communication but support very low data rates.

- **Antennas:** Specialized antenna designs are required, often with large physical dimensions, which can be impractical for many applications.

- **Integration with other Modes:** EM communication is increasingly being considered as part of a hybrid system, where it can provide backup or supplemental communication when acoustic channels are disrupted.

Performance Metrics

- **Range:** EM communication can achieve longer ranges than optical in certain conditions, but effective communication for UUVs is typically limited to short ranges.[53]

- **Bandwidth:** The available bandwidth is very low, often only a few tens or hundreds of bits per second.

- **Latency:** Latency is minimal due to the high speed of EM wave propagation, but overall system performance is constrained by low data throughput.

Maturity of EM Communication

- **Naval:** ELF communication has been used for secure, low-data-rate communication (for hundreds of kilometres) with large submerged vessels (not UUVs) where other methods are impractical.

- **Industrial:** Specialised applications may include data transfer in scenarios where acoustic and optical methods are impractical.

- **Scientific:** Limited applications in scenarios requiring extremely low bandwidth but high reliability over long distances.

---

[53] Dhanak & Xiros, 2016

Comparative Summary

Performance metrics such as range, bandwidth, latency, and energy consumption are key to evaluating underwater communication systems.[54] Table 7-4 and Figure 7-7 summarise key performance aspects, advantages and limitations for the underwater communication technologies discussed.

Table 7-4. Comparison of performance of communication modes

Mode	Range	Bandwidth	Latency	Advantages	Limitations
Acoustic	1 km to >10 km (optimal)	1–20 kbps (typical)	100–1000 ms	Long range; robust in various conditions	Limited bandwidth; high latency due to sound speed
Optical	10–50 m in clear water; <10 m in turbid water	Up to several Mbps	<1 ms (signal speed)	High data rates; low latency	Short range; highly dependent on water clarity
Electromagnetic (ELF/VLF)	Tens to 100s of km (ELF); limited by antenna size	Tens to hundreds of bits/s	Minimal	Long range in very specific applications	Extremely low bandwidth; large antenna require-ments

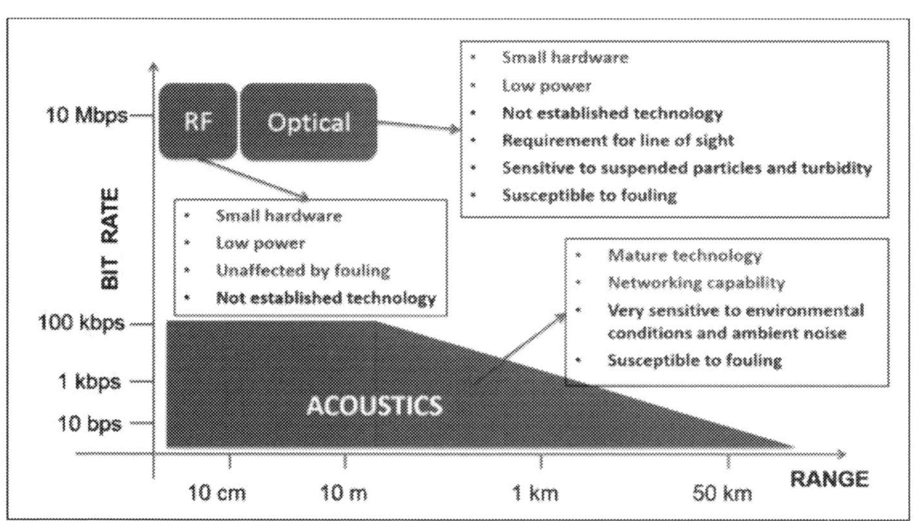

Figure 7-7. Achievable range and data bandwidth of different communication means in the underwater domain[55]

---

54 Stojanovic, 2007; Inzartsev & Pavi, 2009; Kilfoyle & Baggeroer, 2000; Rice & Frater, 2011
55 Ferri, Munafo, Tesei, & Braca, 2017

Emerging Communication Technologies

Driven by interdisciplinary research and the need for higher data throughput, longer ranges, and improved reliability, we look at developments in various technologies that could improve underwater communication in the years ahead.

Hybrid and Multi-Modal Systems

Hybrid communication systems aim to leverage the strengths of multiple modalities to overcome individual limitations.

- **Adaptive Switching:** Some systems are designed to switch between acoustic and optical communication modes based on environmental conditions and data rate requirements. For example, an AUV might use optical communication when in close proximity to a base station to transfer high-bandwidth data and revert to acoustic communication for long-range transmission.[56]

- **Adaptive Multi-Modal Platforms:** Emerging platforms are being designed to dynamically switch between acoustic, optical, and EM channels. These systems use AI-driven decision algorithms to select the optimal channel based on current environmental conditions, operational range, and data rate requirements. Advanced digital signal processing will enable real-time, adaptive switching between communication modes based on environmental factors and mission requirements.[57]

- **Integrated Systems:** By integrating multiple communication modalities, advanced autonomous systems can ensure robust and reliable data transfer under varying conditions. AI-driven algorithms help select the best communication channel based on real-time analysis of environmental factors, required data rate, and range.[58] Development of integrated transceiver units capable of operating across multiple frequency bands will reduce system complexity and power requirements.

- **Collaborative Networking:** Network-centric architectures will allow unmanned vehicles to operate as part of a distributed network, sharing data in a seamless manner even when individual links are compromised.[59]

[56]   Defense Science Board, 2015

[57]   Ghabcheloo, Pascoal, Silvestre, & Kaminer, 2009

[58]   Inzartsev & Pavi, 2009

[59]   Qiu, Zhao, Zhang, Chen, & Chen, 2020

Materials and Transducers

Innovations in materials science are expected to yield improvements in transducer performance.

- **Nanomaterials and Metamaterials:** These materials may offer enhanced acoustic transmission and reception properties, leading to improved sensitivity and bandwidth for acoustic modems.

- **Miniaturized Optical Components:** Advances in laser diodes, photo-detectors, and beam steering technologies could improve the range and reliability of optical communication systems.

- **Robust, Low-Power Electronics:** The development of low-power, high-efficiency electronic components would reduce the energy footprint of underwater communication devices, enabling longer mission durations.

Standardisation and Interoperability

As underwater communication systems evolve, there is a growing need for standardised protocols and interfaces. International bodies and industry consortia are working to establish common standards for underwater data communication (e.g., IEEE and ITU).

Standardised interfaces will enable unmanned systems from different manufacturers to communicate seamlessly, facilitating multi-vendor, multi-domain network-centric operations.[60] These aspects have been covered in the earlier chapter on standards.

Enhanced certification processes will be required to ensure that emerging technologies meet safety, reliability, and performance criteria in the challenging underwater environment.

IoUT and Integration of AI/ML

Underwater autonomous systems of the future will increasingly involve networked sensor and vehicle systems, commonly referred to as Internet of Underwater Things (IoUT) or Subsea Internet of Things (SIoT).[61] The reliability, range and fidelity of underwater communications are crucial for the viability of such networks.[62]

---

[60] Griffiths, 2003

[61] Nkenyereye, Nkenyereye, & Ndibanje, 2024

[62] Gupta & Singh, 2017; Hyland, 2016

- **AI for Communication/Channel Optimization:** AI and machine learning are increasingly applied to optimize underwater communication. These algorithms process sensor data to predict channel conditions, adjust transmission parameters in real time, and manage network traffic among multiple autonomous vehicles. Machine learning algorithms can predict channel conditions (such as ambient noise, multipath interference, and water clarity) and dynamically adjust transmission parameters.[63]

- **Network Management:** Autonomous systems can use AI to manage multi-modal communication networks, deciding in real time which modality to use for data transmission to ensure reliability and energy efficiency.

- **Network-Centric Operations:** Future naval and industrial autonomous systems are expected to operate as part of integrated, network-centric architectures. Advances in underwater communication protocols and middleware (such as DDS and ROS) facilitate the coordination and data sharing necessary for these distributed systems.

- **Data Compression and Error Correction:** AI-driven adaptive algorithms can improve data throughput by optimizing compression schemes and robust error correction tailored to underwater conditions.

- **Energy Efficiency:** Low-power design and energy harvesting techniques (mentioned in the previous chapter) are being integrated with communication systems to extend operational endurance, particularly in long-term environmental monitoring networks.

Conclusion

The current state-of-the-art enables underwater navigation accuracy that is barely sufficient for many scientific and operational missions, and challenges remain in ensuring robustness in highly dynamic and unpredictable environments. Ongoing research into adaptive filtering, real-time processing, and the integration of emerging sensor technologies promise to further reduce drift and improve the overall accuracy and reliability. Improvements in these technologies are critical to enable extended duration missions underwater

---

[63]   Nichols, et al., 2020

and for collaborative (swarm) operations for scientific, industrial, and military underwater missions.

Underwater communication technologies have come a long way since the early days of tethered ROVs. Acoustic communication remains the workhorse for long-range data transfer underwater. However, the range, bandwidth, and reliability remain the limiting factor for operational capability of UUVs in most applications. Even with a combination of acoustic, optical, and electromagnetic communication modalities, achievable ranges and bit rates are barely adequate even at lab scales. Advances in signal processing, sensor integration, and AI-driven adaptive control are hopeful of improving these systems, but they seem to have some distance to go before adequately enabling real-time data transmission, network-centric operations, efficient swarm operations, or operate an underwater IoT.

Future developments are expected to focus on hybrid communication systems that leverage AI for dynamic channel adaptation, the integration of advanced materials for improved transducer performance, and robust, energy-efficient designs that extend operational endurance. Overcoming challenges such as environmental attenuation, limited bandwidth, and latency will require continued research and innovation, supported by interdisciplinary collaboration among academia, industry, and defence organisations.

Underwater navigation and communication technologies are critical enablers and limiters for utilisation of marine autonomous systems. Even as research progresses and new innovations emerge, these technologies will play an increasingly vital as well as limiting role in unlocking the potential of the oceans.

These capabilities form the limiting factors for the ability of UUVs to undertake their intended mission, for which they are fitted with sensors and payloads. Sensors for acoustic communication and navigation have been covered in this chapter. In the next chapter, we look at all types of UUV sensors and payloads, focusing more on aspects not yet covered so far.

References

Alexandris, C., Papageorgas, P., & Piromalis, D. (2024). Positioning Systems for Unmanned Underwater Vehicles: A Comprehensive Review. *Applied Sciences,* 14(21). doi:https://doi.org/10.3390/app14219671

Antonelli, G. (2018). *Underwater Robots.* Springer. doi:10.1007/978-3-319-72124-8

Aoki, T. (2008). Autonomous Underwater Vehicles, Design and Operation. In Ura (ed.), *Urashima. In Autonomous Underwater Vehicles, Design and Operation*, 2nd edition. Ura.

Asai, T., Kojima, J., Asakawa, K., & Iso, T. (2000). Inspection of submarine cable of over 400 km by AUV. *Proceedings of the 2000 International Symposium on Underwater Technology (Cat. No. 00EX418).00*, pp. 133-135. UT; Tokyo, Japan: IEEE. doi:10.1109/ut.2000.852529

Bachmayer, R., Antonelli, E., Fossen, T., Indiveri, G., Yuh, J., & Ødegaard, A. (2008). *Springer Handbook of Robotics*. Heidelberg, D: Springer.

Bahr, A., Leonard, J. J., & Fallon, M. (2009). Cooperative localization for autonomous underwater vehicles. *The International Journal of Robotics Research, 28,* 714–728.

Bhatt, E. C., Viquez, O., & Schmidt, H. (2022). Under-ice acoustic navigation using real-time model-aided range estimation. *Journal of the Acoustical Society of America,* 151(4), 2656–2671. doi:https://doi.org/10.1121/10.0010260

Billon-Coat, A. (ed.). (2018). *Marine Robotics and Applications.* Springer International Publishing AG. doi:10.1007/978-3-319-70724-2

Busby, R. (1976). *Manned Submersibles.* Office of the Oceanographer of the Navy.

Caccia, M. (2006). Vision-based linear motion estimation for unmanned underwater vehicles. *2003 IEEE International Conference on Robotics and Automation (Cat. No. 03CH37422).1,* pp. 977-982. IEEE. doi:10.1109/robot.2003.1241719

Chaudhary, M., Goyal, N., Benslimane, A., Awasthi, L., Alwadain, A., & Singh, A. (2023). Underwater Wireless Sensor Networks: Enabling Technologies for Node Deployment and Data Collection Challenges. *IEEE Internet Things Journal,* 3500–3524.

Choi, H.-T., & Yuh, J. (2016). Underwater Robots. In B. Siciliano, & O. Khatib (eds.), *Springer Handbooks* (pp. 595-622). Heidelberg, D: Springer International Publishing. doi:10.1007/978-3-319-32552-1_25

COCORO (2015, May 27). COCORO: robot swarms use collective cognition to perform tasks. Retrieved from Digital strategy: https://digital-strategy.ec.europa.eu/en/news/cocoro-robot-swarms-use-collective-cognition-perform-tasks.

Defense Science Board. (2015). *Next-Generation Unmanned Undersea Systems.* Tech. rep., Defense Science Board, US Department of Defense.

Demim, F., Benmansour, S., et al. (2021). Simultaneous localisation and mapping for autonomous underwater vehicle using a combined smooth variable structure filter and Extended Kalman Filter. *Journal of Experimental & Theoretical Artificial Intelligence,* 34(4), 621–650. https://doi.org/10.1080/0952813X.2021.1908430

Dhanak, M., & Xiros, N. (2016). *Springer Handbook of Ocean Engineering.* Springer.

DVL1000. (n.d.). Retrieved May 2025, from Nortek: https://www.nortekgroup.com/products/dvl-1000-300m

Eustice, R. M., Singh, H., Leonard, J. J., Walter, M. R., & Ballard, R. D. (2005). Visually navigating the RMS *Titanic* with SLAM information filters. 451–460.

Fenucci, D., Fanelli, F., Consensi, A., Salavasidis, G., Pebody, M., Phillips, A.B. (2024). A multi-platform Guidance, Navigation and Control system for the Autosub family of Autonomous Underwater Vehicles. *Control Engineering Practice.* 146. Retrieved from: https://www.sciencedirect.com/science/article/pii/S0967066124000625

Ferri, G., Munafo, A., Tesei, A., & Braca, P. (2017). Cooperative robotic networks for underwater surveillance: An overview. *IET Radar, Sonar and Navigation.* doi:10.1049/iet-rsn.2017.0074

Finn, A., & Scheding, S. (eds.). (2010). *Developments and Challenges for Autonomous Unmanned Vehicles: A Compendium.* Springer-Verlag Berlin Heidelberg. doi:10.1007/978-3-642-10704-7

Fossen T.I. (2002). *Marine Control Systems: Guidance, Navigation and Control of Ships, Rigs and Underwater Vehicles.* Marine Cybernetics AS, Trondheim

Geyer, R. (1977). *Submersibles and their Use in Oceanography and Ocean Engineering.* New York: Elsevier.

Ghabcheloo, R., Pascoal, A., Silvestre, C., & Kaminer, I. (2009). Coordinated Path Following Control of Multiple Vehicles subject to Bidirectional Communication Constraints. In *Lecture Notes in Control and Information Science* (pp. 93-111). Seoul: Springer-Verlag. doi:10.1007/11505532_6

Goetz, M., Nissen, I., Asterjadhi, A., Casari, P., Zorzi, M., Husøy, T.,& van Walree, P. (2012). *Underwater Acoustic Networking Technologies.* Springer. doi:10.1007/978-1-4471-2926-2

González-García, J., Gómez-Espinosa, A., Cuan-Urquizo, E., García-Valdovinos, L. G., Salgado-Jiménez, T., & Cabello, J. A. (2020). Autonomous Underwater Vehicles: Localization, Navigation, and Communication for Collaborative Missions. *Applied Sciences,* 10(4). doi:https://doi.org/10.3390/app10041256

Griffiths, G. (ed.). (2003). *Technology and Applications of Autonomous Underwater Vehicles.* Taylor & Francis. doi:10.1201/9780203023249

Gupta, S., & Singh, N. (2017). Underwater wireless sensor networks: a review of routing protocols, taxonomy, and future directions. *The Journal of Supercomputing,* 80, 5163-5196. doi:10.1007/s11227-023-05646-w

Hasan, H. (2017). *Underwater Robotics: Science, Design and Fabrication.* UTM Press: Malaysia.

Hegrenæs, Ø., Ramstad, A., Pedersen, T., & Velasco, D. (2016). Validation of a New Generation DVL for Underwater Vehicle Navigation. *IEEE/OES Autonomous Underwater Vehicles (AUV)* (pp. 342-348). Tokyo: IEEE. doi:10.1109/AUV.2016.7778694

Heshmat, M., Saoud, L. S., Abujabal, M., Sultan, A., Elmezain, M., Seneviratne, L., & Hussain, I. (2025). Underwater SLAM Meets Deep Learning: Challenges, Multi-Sensor Integration, and Future Directions. *Sensors, 25*(11). doi:https://doi.org/10.3390/s25113258

Hwangbo, S.-H., Jeon, J.-H., & Park, S.-J. (2013). Wireless Underwater Monitoring Systems Based on Energy Harvestings. *Sensors & Transducers,* 18, 113-119. Retrieved from https://www.researchgate.net/publication/290078574_Wireless_Underwater_Monitoring_Systems_Based_on_Energy_Harvestings/

Hyland, B. (2016, October). Subsea Internet of Things to Transform Amount of Information Available. *Hydro International,* 20. Retrieved from WWW.HYDRO-INTER NATIONAL.COM

Inzartsev, A., & Pavi, A. (2009, January 1). AUV Application for Inspection of Underwater Communications. In A. V. Inzartsev (ed.), *Underwater Vehicles* (pp. 371–398). Rijeka: InTech. doi:10.5772/6704

Kilfoyle, D. B., & Baggeroer, A. B. (2000). The State of the Art in Underwater Acoustic Communication. *IEEE Journal of Oceanic Engineering.*

Kim, H., Lee, Y., & Myung, H. (2020). Factor graph optimization for underwater navigation with acoustic signals. *IEEE Access, 8,* 77720–77731.

Kinsey, J. C., Eustice, R. M., & Whitcomb, L. L. (2006). A survey of underwater vehicle navigation: Recent advances and new challenges. *IFAC Conference of Manoeuvring and Control of Marine Craft.* Lisbon.

Kongsberg Discovery. (2024, September 4). *Smashing AUV records, HUGIN Endurance completes a multi-week fully autonomous mission.* Retrieved May 2025, from Kongsberg.com: https://www.kongsberg.com/newsroom/news-archive/2024/smashing-auv-records-hugin-endurance-completes-a-multi-week-fully-autonomous-mission/

Kongsberg Maritime. (2023). HiPAP and cNODE Positioning Systems. *HiPAP and cNODE Positioning Systems.*

L3Harris. (2022). USBL Acoustic Positioning Systems. *USBL Acoustic Positioning Systems.*

Leonard, J., & Bahr, A. (2016). Autonomous underwater vehicle navigation. In *Springer Handbook of Ocean Engineering* (pp. 341–358). Springer.

Liu, Y., & Bucknall, R. (2019). *Navigation and Control of Autonomous Marine Vehicles.* The Institution of Engineering and Technology. doi:10.1049/pbce019e

McPhail, S. e. (2009). Exploring beneath the ice shelf with Autosub3. *OCEANS 2009 Europe.*

Meduna, Z., Petillot, Y., & Lane, D. (2008). Long-term autonomous underwater vehicle navigation using terrain matching. *IEEE Journal of Oceanic Engineering, 33,* 103–121.

Moore, S., Bohm, H., & Jensen, V. (2010). *Underwater Robotics: Science, Design and Fabrication.* Monterey, CA: Marine Advanced Technology Education Center.

National Research Council, Committee on Autonomous Vehicles in Support of Naval Operations; Naval Studies Board; Division on Engineering and Physical Sciences. (2005). *Autonomous Vehicles in Support of Naval Operations.* The National Academies Press. doi:10.17226/11379

Nichols, R. K., Mumm, H. C., Ryan, J., Lonstein, W., Carter, C., Hood, J. P., & Shay, J. (2020). *Unmanned Vehicle Systems & Operations on Air, Sea, Land.* New Prairie Press. doi:10.4148/2334-0402.1030

Nkenyereye, L., Nkenyereye, L., & Ndibanje, B. (2024). Internet of Underwater Things: A Survey on Simulation Tools and 5G-Based Underwater Networks. *Electronics* (Special Issue: *Artificial Intelligence Empowered Internet of Things*), 13(3). doi:https://doi.org/10.3390/electronics13030474

Partan, J. K. (2007). A Survey of Practical Issues in Underwater Networks. *ACM SIGMOBILE Mobile Computing and Communications Review.*

Paull, L., Saeedi, S., Seto, M., & Li, H. (2014). AUV navigation and localization: A review. *IEEE Journal of Oceanic Engineering, 39(1),* 131–149.

Palomeras N, Carreras M, Andrade-Cetto J. Active SLAM for Autonomous Underwater Exploration. Remote Sensing. 2019; 11(23):2827. Retrieved from: https://doi.org/10.3390/rs11232827

Qiu, T., Zhao, Z., Zhang, T., Chen, C., & Chen, C. (2020). Underwater Internet of Things in Smart Ocean: System Architecture and Open Issues. *IEEE Transactions on Industrial Informatics, 16,* 4297–4307.

RAND Corporation. (2019). *Advancing Autonomous Systems: An Analysis of Current and Future Technology for Unmanned Maritime Vehicles.* Tech. rep., RAND Corporation. doi:10.7249/RR2664

Randeni, S., Schneider, T., Bhatt, E. C., Víquez, O. A., & Schmidt, H. (2022, November). A high-resolution AUV navigation framework with integrated communication and tracking for under-ice deployments. *Journal of Field Robotics*. doi:https://doi.org/10.1002/rob.22133

Rice, J. A., & Frater, N. (2011). Optical Underwater Wireless Communications. *IEEE Journal of Oceanic Engineering*.

Roberts, G. N., & Sutton, R. (eds.). (2013). *Further Advances in Unmanned Marine Vehicles*. The Institution of Engineering and Technology. doi:10.1049/PBCE002E

Stojanovic, M. (2007). Underwater Acoustic Communication: Design Considerations on the Physical Layer. *IEEE Communications Magazine*.

Teixeira, F. C., & Pascoal, A. M. (2007). Geophysical Navigation Of Autonomous Underwater Vehicles. *IFAC Proceedings, 40*(17), 117-122. doi:https://doi.org/10.3182/20070919-3-HR-3904.00022

Tiano, A., & Ferri, G. (2018). Underwater vehicle navigation using SINS, DVL and LBL with intelligent sensor data fusion. *Journal of Intelligent & Robotic Systems, 91*, 263–278.

Tijjani, A. S., & Chemori, A. (2021). *Underwater Vehicles: Design and Applications*. Nova Science Publishers, Inc.

Wynn, R. B., Huvenne, V. A., Bas, T. P., Murton, B. J. (2014, March). Autonomous Underwater Vehicles (AUVs): Their past, present and future contributions to the advancement of marine geoscience. *Marine Geology, 352*, 451-468. doi:http://dx.doi.org/10.1016/j.margeo.2014.03.012

Yu, Z., Zhang, Q., Chen, G., Huang, C., & Lan, J. (2023). Submarine Geophysical Navigation of Autonomous Underwater Vehicle Using Extended Kalman Filter. *2023 IEEE International Conference on Electrical, Automation and Computer Engineering (ICEACE)*. Changchun, China. doi:10.1109/ICEACE60673.2023.10442118

Yuan, X., Martínez, J.-F., Pons, J. S., & Eckert, M. (2017, May). AEKF-SLAM: A new algorithm for robotic underwater navigation. *Sensors*. doi: 10.3390/s17051174

Zhang, S., Zhao, S., An, D., Liu, J., Wang, H., Feng, Y., & Zhao, R. (2022, November). Visual SLAM for underwater vehicles: A survey. *Computer Science Review, 46*. doi:https://doi.org/10.1016/j.cosrev.2022.100510

Zhang, T., Liu, B., & Liu, Y. (2018). Positioning Systems for Jiaolong Deep-Sea Manned Submersible: Sea Trial and Application. *IEEE Access*. doi:10.1109/ACCESS.2018.2881390

8

Sensors and Payloads

UUVs are equipped with an array of sensors and payloads that enable them to collect data about the underwater environment. Sensors are not only the 'eyes and ears' of an AUV but also serve as its scientific instruments, navigation aids, and communication nodes. Steady advances in miniaturisation, energy efficiency, and intelligent data processing of sensors have significantly enhanced the capabilities of AUVs.

We have discussed about sensors as components of AUV design (Chapter 4) and as part of navigation and communication systems (Chapter 7). In this chapter, we focus mainly on equipment characteristics and their integration on the UUV platform.

This chapter reviews the current state of sensor and payload technology on AUVs. The first section provides an overview of sensor types and their purpose or functions on AUVs. Subsequent sections cover key factors that influence sensor selection, analysis of energy and integration aspects, and a view of emerging technologies in underwater sensing. Finally we look at the sensor fit of some typical AUVs as examples.

Overview of Sensors and Payloads

As mentioned in the earlier chapter on AUV design, the sensors and payloads required for the mission objectives should be the starting point of the vehicle design. AUVs and ROVs carry a diverse range of sensors and payloads designed to meet mission-specific objectives. These sensor systems can be grouped broadly into categories that are described below.

Environmental and Oceanographic Sensors

- **Conductivity–Temperature–Depth (CTD) Sensors:** CTD sensors are among the most common payloads on scientific AUVs. They measure the electrical conductivity, temperature, and pressure (depth) of the surrounding water. Data from CTDs are used to calculate salinity and density, both of which are critical for understanding oceanographic processes. Advances in miniaturization now allow CTDs to be integrated into compact packages with lower power consumption while maintaining high accuracy (Figure 8-1).[1]

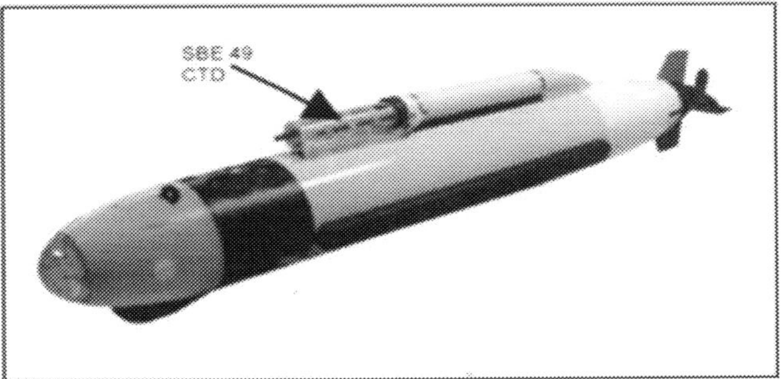

Figure 8-1. CTD Sensor integrated with an AUV[2]

- **Acoustic Doppler Current Profiler (ADCP):** An ADCP measures water current velocities at various depths. It sends sound pulses into the water and then measures the Doppler shift in echoes scattered back by particles suspended in the water. The frequency shift tells the speed and direction of water currents at different depths (i.e., a water column profile).

- **Fluorometers and Turbidity Sensors:** Fluorometers detect chlorophyll concentrations and other fluorescent compounds, providing insights into biological activity and primary productivity. Turbidity sensors measure the clarity of water, which can be used to infer suspended sediment concentrations or detect pollution events.[3]

- **Dissolved Oxygen and pH Sensors:** These sensors provide key chemical data on water quality. Dissolved oxygen sensors help assess

---

[1] https://www.seabird.com/auv-rov-sensors/family; Moore, Bohm, & Jensen, 2010

[2] https://misclab.umeoce.maine.edu/ftp/instruments/CTD%2019+/seabird%20website%2 0disc%20Dec% 202016 /website/applicationnotes/AN93.htm

[3] Freitag, et al., 2005

the habitability of marine environments, while pH sensors are essential for monitoring acidification trends due to climate change.[4]

Acoustic Sensors (Sonars)

Sonars are vital for navigation, mapping, and object detection. They define the basic mission capability of most AUVs, and take primacy in vehicle design.

Side-Scan Sonar (SSS)

Side-scan sonar systems (Figure 8-2) produce images of the seafloor by measuring the intensity of reflected acoustic signals. They are particularly useful for detecting objects and anomalies on the sea floor and detecting objects such as shipwrecks or mines.[5]

Figure 8-2. Edgetech Side Scan Sonar (indicated) on REMUS 100 AUV[6]

Figure 8-3. High resolution side-scan sonar image of a WWII B-25 aircraft discovered on seabed in 2017 in Papua New Guinea
(Source: NOAA[7])

---

[4]   Fletcher, 2001

[5]   Moore, Bohm, & Jensen, 2010

[6]   https://www.researchgate.net/figure/AUV-REMUS100-with-Edgetech-side-scan-sonar_fig3_271190867

[7]   https://oceanexplorer.noaa.gov/technology/sonar/side-scan.html

Side scan sonar is less expensive to run (using an AUV) than, for example, an ROV with a high-definition camera. The resulting 'picture' from side-scan sonar data is made up of dark and light areas. Hard objects protruding from the bottom send a strong echo and create a dark image. Shadows and soft areas, such as mud and sand, send weaker echoes and create light areas. These dark and light images help create accurate maps of the seafloor (Figure 8-3).[8] A lower frequency (50 kilohertz (1 kHz to 100 kHz) can cover large swathes of the seafloor at low image resolution. Higher-frequency pulses (500 kHz to 1 megahertz) record smaller areas but in much greater detail.

Synthetic-Aperture Sonar (SAS)

The SAS provides higher-resolution images by processing multiple pings over time to synthesize a large aperture. It can achieve resolutions down to a few centimetres. A SAS beam is like a 'funnel' that overlaps itself multiple times to achieve highly detailed images (Figure 8-4). As a result, synthetic aperture sonar can map a site at 30 times the resolution of traditional side scan sonar.[9]

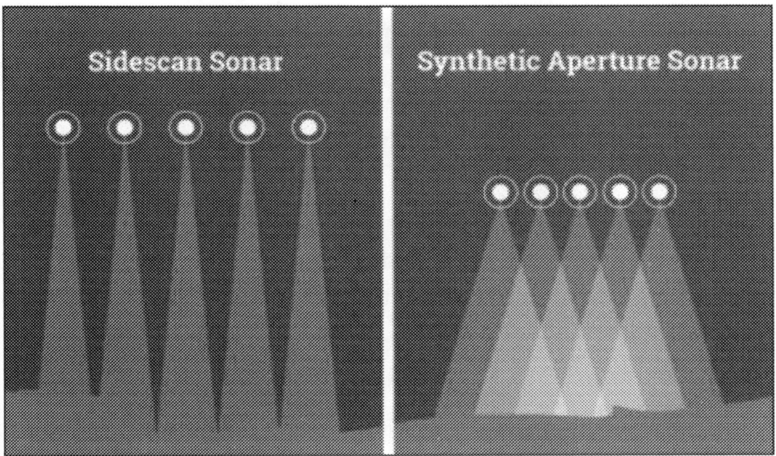

Figure 8-4. Comparison of principle of operation of Side-scan Sonar and Synthetic Aperture Sonar
(Source: NOAA[10])

---

[8] Side Scan Sonar, NOAA

[9] Synthetic Aperture Sonar, NOAA

[10] https://oceanexplorer.noaa.gov/technology/sonar/sas.html

Figure 8-5. Kraken Robotics SAS installed on REMUS AUV[11]

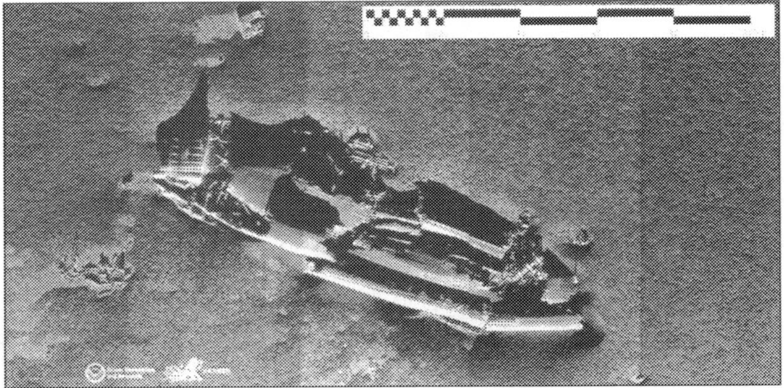

Figure 8-6. The SAS scan of a shipwreck off the coast of Nantucket, USA
(Source: NOAA[12])

Example of an SAS mounted on REMUS AUV is shown in Figure 8-5, and its larger size (compared to the Side scan sonar in Figure 8-2) is apparent. SAS can be valuable for imaging cultural heritage sites like shipwrecks, classifying habitat or biological organisms, and characterizing seafloor sediment makeup (Figure 8-6).

Multibeam Echo Sounder (MBES)

This device emits sound waves in multiple beams to map large swathes of the seafloor. It generates high-resolution detailed bathymetric maps with resolutions typically ranging from 0.5 to 2 metres, depending on frequency and altitude (i.e., height above seabed).

A multibeam sonar uses multiple transducers (in an array) to send and receive sound pulses that map the seafloor or detect other objects. The array

---

11   https://sea-technology.com/tag/synthetic-aperture-sonar

12   *Ibid.*

sends out multiple, simultaneous sonar beams (or sound waves) at once in a fan-shaped pattern. This covers the space both directly under the vessel and out to each side. Multibeam collects two types of data: seafloor depth and backscatter.

The seafloor depth, or bathymetry, is computed by measuring the time it takes for the sound to leave the array, hit the seafloor, and return to the array. Backscatter is a measurement of the intensity of the sound echo that reflects back to the multibeam array. These data are processed to create colourful two- or three- dimensional bathymetric maps to visualize the seafloor (Figure 8-7).[13]

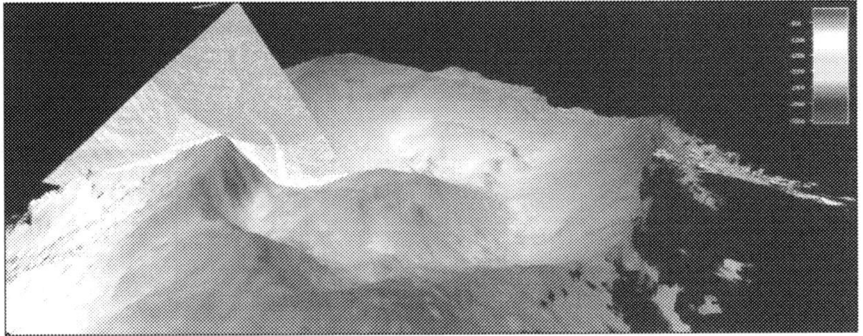

Figure 8-7. A plume of bubbles is seen (in triangle) rising from a seafloor mount, using a single ping of multibeam sonar
(Source: NOAA[14])

Figure 8-8. Multibeam Echosounder (MBES) Surveyor 240-16, weighing 790 grams, size 190x90 mm
(Source: Blue Robotics[15])

---

[13]   Multibeam Sonar, NOAA

[14]   *Ibid.*

[15]   https://bluerobotics.com/store/sonars/echosounders/cerulean-surveyor-240-16-multibeam-echosounder/

Figure 8-9. Multibeam Echosounder (MBES) HydroBeam M4, weighing 5.9 kg, size 230 (Dia) × 175 (H) mm
(Source: Satlab[16])

Multibeam Echosounders find extensive use in hydrographic surveying, offshore oil and gas exploration, oceanography, underwater archaeology, fisheries management, underwater construction, and environmental monitoring. Examples are shown in Figures 8-8 and 8-9. The MBES can be the single most expensive equipment or payload on an AUV.

Doppler Velocity Log (DVL)

A DVL measures the velocity of the vehicle (e.g., AUV or ROV) relative to the seabed or water column. This is crucial for inertial navigation.[17] Similar to ADCP, it uses Doppler shift, but focuses on vehicle motion. It usually has four angled transducers that emit pings downward. If the bottom is in range, it measures bottom-lock velocity. Otherwise, it estimates movement relative to water particles (like an inverted ADCP). Performance aspects have been covered in the previous chapter on underwater navigation and communication.

Sub-Bottom Profiler

A sub-bottom profiler is a a geophysical survey tool that uses sound to map beneath (i.e. into) the seafloor (Figure 8-10).

A sub-bottom profiler emits low-frequency sound pulses that are directed vertically at the seafloor via a transducer. When some of these pulses hit the substrate, they may penetrate initial layers of seafloor before reaching horizons of various sediment types. How far the pulses penetrate, and how long they

---

[16]   https://www.satlab.com.se/wp-content/uploads/2025/04/HydroBeam-M4-Brochure-EN-20250422s.pdf

[17]   Hegrenæs, Ramstad, Pedersen, & Velasco, 2016

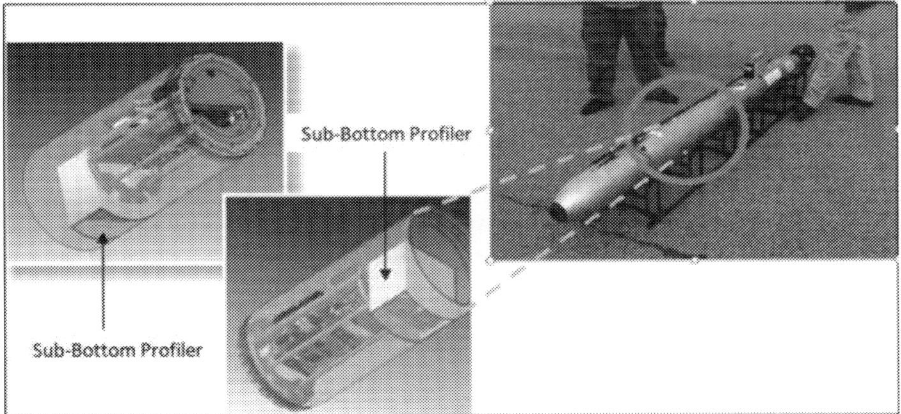

Figure 8-10. Teledyne 'Benthos III Chirp' Sub-Bottom Profiler fitted inside a GaviaScientific AUV[18]

take to bounce back, will vary based on the makeup of the substrate. By calculating the time taken for a sound pulse to return to the ship, the sediment and layer composition, thickness, and other characteristics are determined.

It is thus possible to generate cross sections or vertical slices depicting sediments below the seafloor. Through interpreting these cross sections, hydrographers can identify features such as volcanic ridges, underwater landslides, ancient river beds, buried objects, and other geological features.[19]

Optical Sensors

- **Cameras and Imaging Systems:** High-definition digital cameras (including low-light and hyper-spectral cameras) are increasingly being integrated on AUVs for visual imaging of marine habitats, infrastructure inspection, and object recognition. Advances in CMOS sensor technology and low-light sensitivity have enabled higher-resolution imaging with lower power consumption.[20]

- **LiDAR and Laser Scanning:** Although less common due to limited range in turbid water, underwater LiDAR systems are emerging to provide high-resolution 3D mapping of seafloor structures and complex underwater objects.

---

[18]   https://www.oceanbytes.org/2011/04/01/sub-bottom-profiling-using-an-auv/

[19]   Sub-Bottom Profiler, 2023

[20]   Caccia, 2006

Chemical and Biological Sensors

- **Chemical Analysers:** These sensors can detect dissolved chemicals, nutrients (e.g., nitrate, phosphate), and contaminants. They are often miniaturized for AUV integration, enabling real-time water quality monitoring.

- **Biosensors:** Biosensors, such as those that measure chlorophyll fluorescence or detect microbial activity, help in ecological studies and can be used to monitor harmful algal blooms.

- **Radiometric Sensors:** Radiometers measure light intensity and spectral distribution in underwater environments. They are critical in studying the optical properties of water and in calibrating imaging systems.

Navigation and Positioning Sensors

- **Inertial Measurement Units (IMUs):** IMUs combine accelerometers, gyroscopes, and sometimes magnetometers to provide data on the AUV's orientation and movement. They are essential for dead-reckoning navigation when GPS is unavailable underwater.[21] Even a miniature ultra-high accuracy MEMS IMU (example shown in Figure 8-11) can provide inertial performance exceeding that of even some FOG IMUs. A typical model features a bias stability of 0.4 degrees per hour.

Figure 8-11. Motus MEMS IMU weighing 26 grams[22]

- **Acoustic Positioning Systems:** Systems such as USBL (Ultra-Short Baseline) and LBL (Long Baseline) allow the AUV to determine its position relative to known acoustic beacons, thereby supplementing inertial navigation.[23] These can also be categorised under acoustic sensors. They aid in collaborative missions among multiple AUVs. These have been discussed in the earlier chapter on underwater navigation and communication.

---

[21] Tiano & Ferri, 2018

[22] https://www.unmannedsystemstechnology.com/feature/selecting-an-inertial-measurement-unit-imu-for-uav-applications/

[23] Stojanovic, 2007

Communication Equipment

- **Acoustic Modem:** Acoustic modems are critical for underwater wireless communication. It encodes and sends digital data (e.g., telemetry, commands, or sensor readings) in the form of acoustic signals. Its receiver decodes the signals into usable information. Communication performance of typical models were mentioned in Chapter 7 (underwater communication). Typical equipment features for integration on UUV are summarised in Table 8-1 and shown in Figure 8-12.

Table 8-1. Characteristics of Acoustic Modems for AUVs[24]

Parameter	*Typical Range/Values*	*Remarks*
Size	20–50 cm length, 7–15 cm diameter	Compact for AUV integration; larger models exist for long-range or high-power use
Weight	1–5 kg in air	Lighter in water due to buoyancy
Power Consumption	Transmit: 5–50 W Receive/Idle: 0.5–2 W	Depends on range and data rate; low-power sleep modes often available
Operating Voltage	12–36 V DC	For compatibility with various platforms
Frequency Bands	8–50 kHz (most common)	Lower freq = longer range, lower bandwidth; Higher freq = higher bandwidth, shorter range
Data Rate (Bandwidth)	**Short range:** up to 50–100 kbps; **Long range (5–10 km):** ~100–1000 bps	Bandwidth heavily depends on range, environment, and modulation
Communication Range	500 m – 10+ km	Line-of-sight; affected by salinity, noise, multipath, etc.
Latency	0.5–2 s typical round-trip	Due to sound speed in water (~1500 m/s) and processing delay
Bit Error Rate (BER)	10^{-5} to 10^{-8} (with error correction)	Varies with SNR and channel conditions
Supported Protocols	Proprietary, or standards like JANUS (NATO), WHOI Micro-Modem protocol	Some modems allow networking with TDMA/CSMA/Token-passing
Duplex Mode	Half-duplex (mostly)	Full-duplex is rare and power-hungry
Depth Rating	100–6000 m	Based on housing material (aluminium, titanium, etc.)

---

[24] Hwangbo, Jeon, & Park, 2013

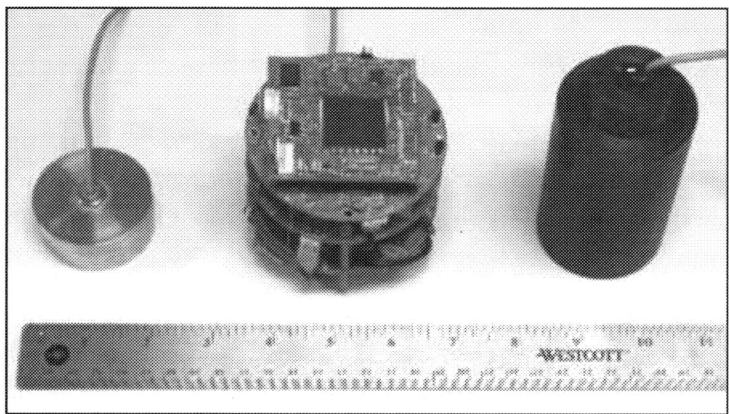

**Figure 8-12. Acoustic modem (middle) with directional (left)
and omni-directional transducer (right)[25]**

- **GPS Antenna:** For UUVs, GPS antennas are crucial for surface navigation and positioning when the vehicle surfaces. They are often used in conjunction with other underwater positioning systems like acoustic modems or inertial navigation systems (INS). These antennas need to be rugged, waterproof, depth-rated, and capable of operating in harsh marine environments.[26]

- **RF Antenna:** AUVs and UUVs often utilize UHF (225-400 MHz) for LOS communication and potentially L-band or S-band for SATCOM. Due to the limited space within AUVs and UUVs, compact antennas are essential. The antenna must withstand the pressure, salinity, and potential impacts associated with underwater operations. UHF antennas are commonly used for LOS communication between the AUV/UUV and a surface vessel or base station. L-band or S-band antennas enable communication with satellites, facilitating long-range communication and data transfer. Antennas that support mesh networking can facilitate communication between multiple AUVs or UUVs, enabling collaborative operations.[27] Some antennas are designed to receive GPS signals while the UUV is at shallow depth, enabling accurate positioning.

---

[25] https://www.researchgate.net/figure/Acoustic-modem-middle-with-directional-left-and-omnidirectional-transducer-right_fig1_329617107

[26] Alexandris, Papageorgas, & Piromalis, 2024

[27] Zhou, 2021

Payload Integration and Multi-Sensor Suites

Modern AUVs are increasingly designed with modular payload bays that can accommodate different combinations of sensors based on mission requirements. For instance, an AUV deployed for mine countermeasures might integrate synthetic-aperture sonar with magnetometers and high-resolution video cameras, while another used for oceanographic sampling might carry a suite of CTD sensors, fluorometers, and water samplers. The flexibility in the basic design of the AUV for integrating various options for payload is a key advantage, allowing the same platform to be reconfigured for diverse missions.[28] A typical arrangement is shown in Figure 8-13.

Factors Affecting Sensor Selection

Sensor selection for UUVs must balance performance requirements, environmental conditions, integration constraints, and mission objectives. These factors are listed further to serve as a ready reckoner in making choices about the sensor fit, and between different models of a type of sensor.

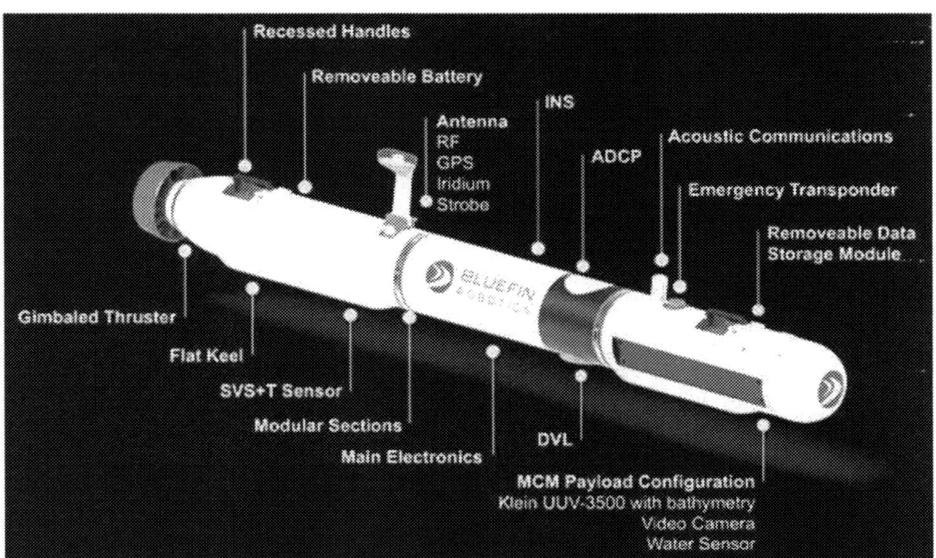

Figure 8-13. Arrangement of sensors and other components on Bluefin-9M AUV
(Source: General Dynamics)[29]

---

[28] Allen, et al., 2002; Chocron, Chemori, Creuze, & Lozano, 2017

[29] https://gdmissionsystems.com/discontinued-products/bluefin-9-m-autonomous-under water-vehicle

Mission Requirements

The primary driver of sensor selection is the mission profile, which could be any (or all) of the following:

- **Scientific Research:** High-precision measurements, multi-parameter data collection, and long-duration deployments demand sensors with high resolution and stability.

- **Military and Defence:** Robustness, stealth, and the ability to detect and classify objects (e.g., mines, submarines) are paramount.

- **Industrial Inspections:** Sensors used in offshore pipeline monitoring or infrastructure inspections need to be rugged, reliable, and capable of delivering real-time data.

Mission requirements determine the type of data needed (e.g., acoustic imaging or optical imaging), the resolution of that data, and the acceptable error margins. In addition, mission duration, area coverage, and the need for autonomous decision-making influence sensor selection.[30]

Environmental Conditions

Sensors must be selected based on the environment in which the AUV will operate. Some of the important factors are listed below:

- **Depth and Pressure:** Deep-sea missions require sensors that can withstand high hydrostatic pressures. Pressure-tolerant or pressure-compensated designs are essential.

- **Temperature:** Extreme cold or warm conditions can affect sensor performance. Components must be able to operate over the anticipated temperature range.

- **Water Clarity and Turbidity:** Optical sensors are highly sensitive to water clarity. In turbid environments, sensors with enhanced low-light performance or alternative imaging modalities (e.g., sonar) may be necessary.

- **Salinity and Corrosiveness:** Saltwater is highly corrosive, so sensor materials and housings must be selected for durability and resistance to biofouling.

---

[30] Griffiths, 2003

Power Consumption and Energy Budget

AUVs are energy-constrained platforms. Sensor power requirements must be balanced against the overall energy budget of the vehicle.[31]

- **Low-Power Design:** Many sensors are designed to operate at low power (ranging from a few milliwatts to a few watts). However, high-resolution imaging systems or active sonar modules may require significantly more power (from 5 W to tens of watts).

- **Duty Cycle and Operational Modes:** Sensors that do not need to operate continuously can be scheduled to power on only when needed, reducing overall energy consumption.

- **Integration with Energy-Harvesting or in situ Charging:** Emerging technologies such as underwater wireless power transfer or energy harvesting (e.g., from wave energy) can supplement battery power, particularly for long-duration missions.

Weight, Size, and Space Constraints

Due to the limited payload capacity and space within an AUV, sensor size and weight are critical considerations.

Miniature size is always an advantage from the platform design viewpoint. Advances in MEMS (micro-electro-mechanical systems) and CMOS sensor technology have led to smaller, lighter sensors that still maintain high performance.

Sensor packages must be designed to fit within the available payload bays while leaving room for other critical components (e.g., batteries, communication systems). Thus, integration into the vehicle must be compact.

A modular approach allows different sensor suites to be swapped in and out without major modifications to the AUV's overall architecture. Many modern AUVs, such as the REMUS series, incorporate plug-and-play payload modules that optimize available space.[32]

Data Bandwidth and Processing Requirements

Sensors that generate large volumes of data (such as high-definition video or synthetic-aperture sonar) impose significant demands on onboard storage and processing capabilities.[33]

---

[31] Defense Science Board, 2015

[32] REMUS UUVs

[33] Choi & Yuh, 2016

- **Real-Time Processing:** In missions where immediate data interpretation is crucial (e.g., object detection in mine countermeasures), sensors must be paired with powerful onboard processors that can analyse data in real time.

- **Compression and Storage:** High-data-rate sensors require efficient compression algorithms to store and transmit data without overwhelming the AUV's communication links.

- **Communication Bandwidth:** Underwater communication channels (typically acoustic) have limited bandwidth. Sensors must be selected and configured so that data output is manageable given these constraints.

Reliability, Robustness, and Maintenance

The harsh underwater environment necessitates sensors that are robust and require minimal maintenance. Relevant considerations are listed:

- **Corrosion Resistance:** Materials and enclosures must resist corrosion from saltwater.

- **Biofouling Prevention:** Sensors deployed for long durations may suffer from biofouling, which can degrade performance. Anti-fouling coatings and mechanical cleaning systems are sometimes employed.

- **Long-Term Stability:** Sensors must maintain their calibration and accuracy over extended missions without the need for frequent recalibration.

- **Ease of Replacement and Repair:** In modular systems, sensors should be easily replaceable to reduce downtime during maintenance. Standardisation of components and compatibility of interfaces (physical as well as communication) are highly significant considerations.

Cost and Economic Considerations

Finally, the cost of sensors can be a limiting factor, particularly for research projects or large-scale deployment.

- **Cost-Performance Trade-Off:** While high-end sensors may offer superior accuracy and resolution, they significantly affect project cost. Engineers must balance performance requirements with available budgets.

- **Scalability:** For applications involving large networks, such as the Subsea Internet of Things (SIoT) or Underwater Wireless Sensor Networks (UWSNs), cost-effective, scalable sensor solutions are preferred.

- **Customisation:** The extent of customisation feasible in a particular sensor may determine its utility in a specific operating environment. Performance characteristics are generally advertised for ideal conditions and modifications may be necessary at actual site. The necessity of the equipment supplier's representative (and the implicit dependence) for such tuning needs to be carefully evaluated.

Typical Energy, Power, Weight and Space

For designing an effective payload suite and an efficient UUV around it, specifications of sensors form basic inputs. Although figures vary based on sensor type and manufacturer, typical and indicative values are given in this section for providing the UUV design team a feel for relevant numbers.

Energy and Power Requirements

Since AUVs are battery-operated with finite energy capacity, sensors are selected for minimising power consumption and their duty cycles are managed carefully to conserve power.

- **Low-Power Sensors:** Many environmental sensors (e.g., CTDs, temperature, conductivity) operate on the order of milliwatts (typically 50 mW to 500 mW). Their design emphasizes energy efficiency to allow long-term deployment. Chemical/biological sensors typically range from 50 mW to 1 W, depending on the sensor type and duty cycle.

- **Active Sonar Systems:** Sonar sensors, particularly those with active transmission (such as synthetic-aperture sonar or multibeam sonars), may require power levels ranging from 5 W to 50 W, depending on pulse power, beam-forming requirements, and resolution. Active sonar (e.g., SAS, Multibeam) can require between 5 W and 50 W when transmitting pulses. They are often operated intermittently to conserve energy. Passive sonar (e.g., hydrophones) usually consume less power (approximately 1–5 W).[34]

---

[34] Aoki, 2008

- **Optical Imaging Systems:** High-definition video cameras or hyper-spectral imagers can range from 1-5 W for low-resolution sensors up to 10–20 W for high-end, high-resolution systems, particularly when processing is done onboard.

- **Data Processing and Onboard Computing:** Depending on the complexity of the processing (e.g., real-time object recognition, adaptive control algorithms), onboard computers may require about 5 W to 30 W.

- **Total Payload Power:** In many integrated AUV payload suites, the total power requirement may range from 10 W to 100 W. For missions requiring high-resolution imaging or high-data-rate acoustic systems, power consumption may exceed 100 W.[35]

Weight and Space Considerations

Advances in sensor miniaturization have enabled many sensors to be produced in compact packages. However, sonar and communication sensors may occupy significant space relative to vehicle size and may also govern the vehicle configuration. Salient considerations and typical values are given here.[36]

- **Imaging Sensors:** High-definition cameras used on AUVs generally weigh between 0.5 and 2 kg. When combined with additional optics and housing for underwater use, a complete imaging system might range from 1 to 3 kg. These systems are designed to be compact, often occupying less than 2 litres.

- **Sonar Systems:** Acoustic transducers and sonar modules vary in size. Small single-beam sonars can weigh a few hundred grams, while multibeam or synthetic-aperture sonar arrays may weigh several kilograms and require larger physical housings. They often require a dedicated mounting structure, with typical volumes in the range of 1–5 litres.

- **Chemical and Biological Sensors:** Many of these sensors are designed using microfluidic techniques and MEMS technology. They can be quite lightweight (often less than 500 g) and compact (typically a few cubic decimetres). A modern CTD sensor would weigh less than 1 kg and occupy a volume of 0.5–1 Litre.

---

[35]   Bachmayer, et al., 2008

[36]   Tijjani & Chemori, 2021

- **Navigation Sensors (IMUs, DVLs, USBL modules):** Weight can vary from 0.5 kg (for simple IMUs) to 2–3 kg for integrated navigation packages. These are typically designed for minimal footprint and can be integrated into the vehicle's hull or internal frame.

Payload Integration

AUV platforms usually have strict space and weight limits. Smaller AUVs (such as the REMUS 100) may have a payload capacity of only a few kilograms, whereas larger AUVs (like the REMUS 6000) can carry payloads of tens of kilograms.[37]

Sensors must be mounted securely and oriented correctly (Figure 8-14). The integration may involve custom-designed mounting brackets, waterproof housings, and shock absorbers. Designers strive for a balance between robust sensor placement and minimal interference with the hydrodynamic profile of the vehicle.[38]

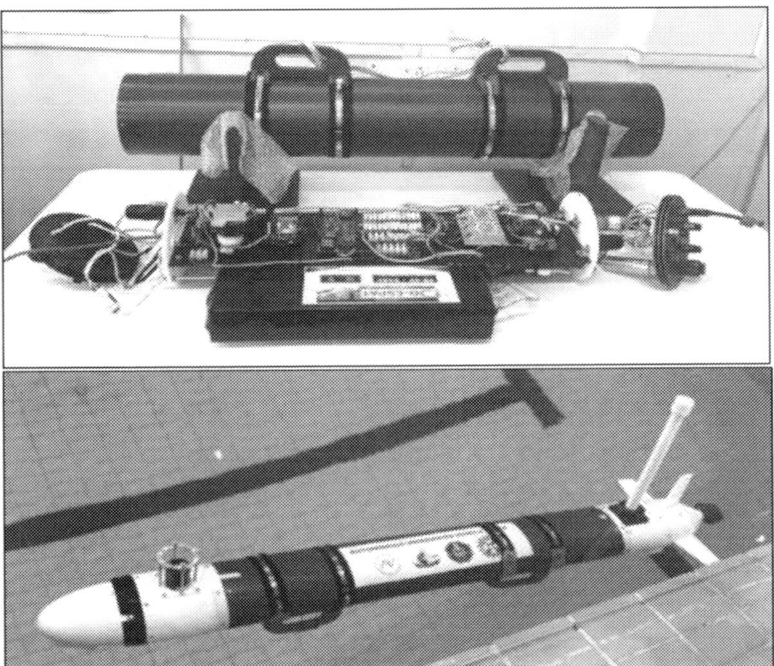

Figure 8-14. AUV Jaya before and after integration
(Source: Author/NIO Goa)

---

[37] Billon-Coat, 2018
[38] Rutherford, 2008

Modular design strategies such as a mission-specific payload bay may provide design flexibility, but there may be a trade-off in the effective use of space possible by stacking or integrating sensor packages.

Sensors that need to be exposed to sea water usually come in pressure-resistant housing and need to be interfaced and calibrated after integration on the vehicle.

Typical Payload Specifications

For a typical mid-range AUV used for environmental and geophysical surveys, a representative sensor payload might include the following:

- **CTD Sensor:** Weighing ~0.8 kg, occupying 1 litre, with a power consumption of 200 mW.

- **Multibeam Sonar:** A compact multibeam sonar module weighing ~3 kg, requiring 30 W during operation.

- **High-Definition Camera System:** A digital imaging package weighing ~1.5 kg, consuming 5 W continuously.

- **Doppler Velocity Log (DVL):** Weighing ~2 kg, with power consumption around 10 W.

- **Additional Chemical Sensors (e.g., dissolved oxygen, pH, nitrate):** Combined weight of ~1 kg, power consumption of ~500 mW each (with duty cycling, average consumption might be lower).

For this configuration, the total payload might weigh approximately 8–10 kg and require an average power budget of roughly 50–60 W when all sensors are operating continuously. In practice, many sensors do not run continuously (they may be duty cycled), so the overall energy consumption is reduced.

Integration constraints, including available payload bay volume and mass limits, drive design trade-offs between sensor quality, size, and energy use.

Emerging Developments

As AUV missions become more demanding and the need for real-time, high-resolution underwater data increases, sensor technology is evolving rapidly. Some of the emerging and futuristic trends are outlined here.

Miniaturized Multi-Parameter Sensor Systems

Developments in MEMS, microfluidics and complementary metal–oxide–semiconductor (CMOS) technologies are enabling the production of integrated multi-parameter sensors that can measure several environmental variables simultaneously.[39] These 'lab-on-a-chip' devices can combine:

- Temperature, conductivity, and depth measurements (CTD)
- Chemical sensing (e.g., pH, dissolved oxygen, nitrates, and heavy metals)
- Biological parameters (e.g., chlorophyll fluorescence for phytoplankton)

These systems not only reduce size and weight but also simplify the integration process by providing a single, unified sensor package. New materials and manufacturing techniques allow for ultra-compact sensor housings that are robust against pressure and corrosion. For instance, composite materials and advanced polymer coatings are being used to protect delicate sensor elements while minimizing weight and size.

Smart and Adaptive Sensors

Future sensors are expected to incorporate onboard processing and machine-learning algorithms, with advanced features.[40]

- **Self-Calibration:** Sensors that automatically calibrate themselves to account for drift or environmental changes. Advanced optical sensors now incorporate algorithms that adjust for lens distortion, chromatic aberration, and backscatter in real time, improving data reliability over long-duration missions.

- **Adaptive Sampling:** Sensors that dynamically adjust their sampling rates based on real-time environmental variability or mission-critical events. This adaptive behaviour allows an AUV to conserve energy during periods of low variability and increase data acquisition during events such as phytoplankton blooms or pollutant dispersal.

- **Embedded AI:** Integration of AI to perform preliminary data analysis onboard, reducing the amount of data that must be transmitted and allowing for autonomous decision-making. By incorporating artificial intelligence (AI) algorithms into sensor systems, AUVs can perform real-time data analysis and make autonomous decisions about sensor

---

[39] Carder, et al., 2001
[40] Antonelli, 2018

activation, data compression, and anomaly detection. For example, machine-learning-based image enhancement techniques can automatically correct for underwater colour distortion and blur.

High-Resolution Imaging

- **Synthetic Aperture Sonar:** Synthetic-Aperture Sonar (SAS) systems continue to evolve, providing finer resolution and improved image quality for seabed mapping, mine detection, and infrastructure inspection. The integration of SAS with AUV platforms is enabling detailed 3D reconstructions of underwater environments.[41]

- **Optical CMOS sensors:.** New CMOS sensors with enhanced low-light performance and higher dynamic range allow for better image quality even in the murky underwater environment.

- **Hyper-spectral Imaging:** Optical sensors that capture a wide spectrum of light wavelengths are emerging, providing detailed spectral signatures that could be used for environmental monitoring, habitat classification, and even chemical detection.

- **Underwater LiDAR:** Although challenging because of scattering effects, advances in laser technology and signal processing could eventually make underwater LiDAR feasible for 3D mapping in clear water.[42]

Hybrid Acoustic-Optical Sensors

Hybrid sensors can offer complementary data, wherein acoustic sensors provide robust range and imaging in turbid waters, while optical sensors deliver high-resolution imagery in clearer conditions.[43] To overcome the limitations of individual modalities, researchers are exploring hybrid sensors that integrate both acoustic and optical components.

- **Dual-Mode Imaging:** Such sensors could operate in both optical and acoustic modes, switching between them depending on water clarity and ambient light conditions.[44]

---

[41] Wadoo & Kachroo, 2011

[42] Rice & Frater, Optical Underwater Wireless Communications, 2011; Roberts & Sutton, 2013.

[43] Balasuriya & Ura, 2002

[44] Yuan, Martínez, Pons, & Eckert, 2017; Zhang, et al., 2022

- **Sensor Fusion:** Combining data from acoustic and optical sensors can yield a more complete understanding of the environment. Advanced sensor fusion algorithms are being developed to integrate disparate data types into coherent models of the seafloor and water column.[45]

- **Extensible Payload Modules ('Whiskers'):** These modules are designed for easy integration into AUVs, offering the flexibility to combine optical, acoustic, and chemical sensors into a single payload package. Such modules emphasize rapid reconfiguration and adaptive mission planning.

Biomimetic and Bio-Inspired Sensors

Taking inspiration from marine organisms, researchers are developing sensors that mimic biological systems.[46]

- **Biomimetic Optical Sensors:** Inspired by the eyes of deep-sea creatures, these sensors aim to achieve superior low-light performance and wide dynamic ranges.

- **Acoustic Mimicry:** Sensors that mimic the hearing mechanisms of marine mammals could lead to improved detection capabilities in noisy environments.

- **Geophysical Sensors:** Sensors for gravimetry or magnetometry may become viable in the future.[47]

- **Self-Healing Materials:** Emerging research into self-healing polymers and coatings could extend the lifespan of sensor housings by automatically repairing minor damage caused by pressure or biofouling. This could extend sensor lifespan and reduce maintenance.

Payloads on Typical AUVs

Several AUV platforms can serve as case studies of how sensor payloads are integrated and utilized. A few widely used AUV designs are discussed here.

---

[45] Cecchi, Caiti, Fioravanti, Baralli, & Bovio, 2005

[46] Hasan, 2017

[47] Teixeira & Pascoal, 2007

Bluefin

The Bluefin-21 AUV (Figure 8-15) is a well-known platform used in high-profile missions such as the search for Malaysia Airlines Flight 370.[48]

Figure 8-15. Bluefin-21 Macrura UUV deployed in ICEX 2020
(Source: General Dynamics[49])

Its payload typically includes:

- **Acoustic Sensors:** An EdgeTech 2200-M sonar, a sub-bottom profiler, and an echo sounder to enable detailed mapping of the seafloor.

- **Optical Imaging:** A Prosilica high-resolution camera is used for visual documentation.

- **Navigation:** An inertial navigation system (INS) and an ultra-short baseline (USBL) positioning system provide accurate location data.

- **Energy Considerations:** Bluefin-21 is powered by nine lithium-polymer battery packs (each rated at 1.5 kWh), providing a total energy capacity of approximately 13.5 kWh. This design balances payload weight (typically a few kilograms per sensor module) with the energy required for long-duration missions.

This integrated payload allows Bluefin-21 to conduct high-resolution sonar mapping and detailed visual inspections over extended missions.

---

[48] Bluefin Robotics Unmanned Underwater Vehicles
[49] *Ibid.*

REMUS

The REMUS family of AUVs (Figure 8-16), developed in collaboration with institutions like Woods Hole Oceanographic Institution (WHOI), is known for its modular payload design.[50]

Figure 8-16. REMUS UUV
(Source: HHI[51])

The REMUS series illustrates how sensor payloads are suited for mission requirements, from detailed seabed mapping in the REMUS 6000 to high-frequency environmental monitoring in the REMUS 100.

- **REMUS 6000:** This large vehicle can carry multiple payloads including multibeam sonars, side-scan sonars, CTDs, and high-definition cameras. Its modular design enables rapid swapping of sensor packages.

- **REMUS 300 and REMUS 100:** These smaller platforms typically integrate essential navigation sensors (IMU, DVL, pressure sensors) along with basic imaging and environmental sensors for scientific surveys. Their smaller payload bays demand miniaturized, low-power sensors.[52]

---

[50] REMUS AUV, WHOI

[51] REMUS UUVs, HHI

[52] Corfield & Hillenbrand, 2003

Sentry

'Sentry' is an AUV designed by the Woods Hole Oceanographic Institute (WHOI) for deep-sea research (Figure 8-17). Sentry's sensors are integrated into a system designed for deep deployments (up to 6,000 m), and its payload modules are optimized for low power consumption to extend mission duration. It has a Doppler velocity log (DVL) and inertial navigation systems for positioning. Environmental and imaging sensors include CTD sensors and digital cameras. It can carry additional instruments such as magnetometers and chemical analysers.

Figure 8-17. Sentry AUV developed by WHOI[53]

Sentry exemplifies the integration of high-precision, deep-sea sensors into a compact AUV platform with a novel form designed for extreme conditions.[54]

Conclusion

Sensors and payloads carried by Autonomous Underwater Vehicles form the core of their operational capabilities. Today's AUVs integrate a wide variety of sensors, ranging from environmental and acoustic to optical and chemical, in order to perform diverse missions of deep-sea exploration, military surveillance, and environmental monitoring.

Sensor selection for AUVs is driven by multiple factors: mission objectives, environmental conditions, power and energy constraints, weight and space

---

[53]   Sentry, WHOI
[54]   *Ibid.*

limitations, data processing needs, and overall cost. In many cases, trade-offs must be made to balance performance with efficiency and reliability. The typical specifications and suggestions provided in this chapter are meant to serve as a starting point for AUV designers, who must often customise sensor configurations based on the specific vehicle size and mission parameters.

Recent developments in sensor technology are opening new possibilities, such as advances in miniaturization, smart sensor design, low-power electronics, and hybrid sensor modalities.[55] Emerging trends such as adaptive, self-calibrating sensors, energy harvesting techniques, and the integration of the Internet of Underwater Things (IoUT) appear promising.[56]

Integration of these sensors and their functioning within an AUV are linked to the all-essential feature of vehicle autonomy. The next chapter discusses autonomy, control algorithms and operator (human) interfaces.

References

Alexandris, C., Papageorgas, P., & Piromalis, D. (2024). Positioning Systems for Unmanned Underwater Vehicles: A Comprehensive Review. *Applied Sciences,* 14(21). doi:https://doi.org/10.3390/app14219671

Allen, B., Stokey, R., Austin, T., Forrester, N., Goldsborough, R., Purcell, M., & Von Alt, C. (2002). REMUS: a small, low cost AUV; system description, field trials and performance results. *Oceans '97. MTS/IEEE Conference Proceedings* 2, pp. 994-1000. IEEE. doi:10.1109/oceans.1997.624126

Antonelli, G. (2018). *Underwater Robots.* Springer. doi:10.1007/978-3-319-72124-8

Aoki, T. (2008). Autonomous Underwater Vehicles, Design and Operation. In Ura (ed.), *Urashima. In Autonomous Underwater Vehicles, Design and Operation,* 2nd Edition.

AUV ROV Sensors. (n.d.). Retrieved May 2025, from Sea-Bird Scientific: https://www.seabird.com/auv-rov-sensors/family

Bachmayer, R., Antonelli, E., Fossen, T., Indiveri, G., Yuh, J., & Ødegaard, A. (2008). *Springer Handbook of Robotics.* Heidelberg, D: Springer.

Balasuriya, A., & Ura, T. (2002). Autonomous target tracking by underwater robots based on vision. *Proceedings of 1998 International Symposium on Underwater Technology* 29, pp. 1351–1367. IEEE. doi:10.1109/ut.1998.670089

Billon-Coat, A. (ed.). (2018). *Marine Robotics and Applications.* Springer International Publishing AG. doi:10.1007/978-3-319-70724-2

Bluefin Robotics Unmanned Underwater Vehicles. (n.d.). Retrieved May 2025, from General Dynamics Mission Systems: https://gdmissionsystems.com/underwater-vehicles/bluefin-robotics

---

[55]   Martin, et al., 2019

[56]   Kilfoyle & Baggeroer, 2000

Caccia, M. (2006). Vision-based linear motion estimation for unmanned underwater vehicles. *2003 IEEE International Conference on Robotics and Automation (Cat. No.03CH37422).1*, pp. 977-982. IEEE. doi:10.1109/robot.2003.1241719

Carder, K., Costello, D., Warrior, H., Langebrake, L., Hou, W., Patten, J., & Kaltenbacher, E. (2001, October). Ocean-science mission needs: real-time AUV data for command, control, and model inputs [West Florida Shelf]. *IEEE Journal of Oceanic Engineering*, 26, 742-751. doi:10.1109/48.972116

Cecchi, D., Caiti, A., Fioravanti, S., Baralli, F., & Bovio, E. (2005). Target detection using multiple autonomous underwater vehicles. *Proceedings IARP International Workshop on Under-water Robotics*, (pp. 161–168). Genova.

Chaudhary, M., Goyal, N., Benslimane, A., Awasthi, L., Alwadain, A., & Singh, A. (2023). Underwater Wireless Sensor Networks: Enabling Technologies for Node Deployment and Data Collection Challenges. *IEEE Internet Things Journal*, 3500–3524.

Chocron, O., Chemori, A., Creuze, V., & Lozano, R. (2017). Design and Control of Underwater Vehicles. In L. Jaulin, E. Maître, & J. Yuh (eds.), *Marine Robotics and Applications* (pp. 121–145). Cham: Springer International Publishing.

Choi, H.-T., & Yuh, J. (2016). Underwater Robots. In B. Siciliano, & O. Khatib (eds.), *Springer Handbooks* (pp. 595-622). Heidelberg, D: Springer International Publishing. doi:10.1007/978-3-319-32552-1_25

Corfield, S., & Hillenbrand, C. (2003). Defence Applications for Unmanned Underwater Vehicles. In G. Griffiths (ed.), *Technology and Applications of Autonomous Underwater Vehicles* (pp. 161-178). London: CRC Press. doi:10.1201/9780203522301.ch10

Defense Science Board. (2015). *Next-Generation Unmanned Undersea Systems*. Tech. rep., Defense Science Board, U.S. Department of Defense.

Finn, A., & Scheding, S. (eds.). (2010). *Developments and Challenges for Autonomous Unmanned Vehicles: A Compendium*. Springer-Verlag Berlin Heidelberg. doi:10.1007/978-3-642-10704-7

Fletcher, B. (2001). Chemical plume mapping with an autonomous underwater vehicle. *MTS/IEEE Oceans 2001. An Ocean Odyssey. Conference Proceedings (IEEE Cat. No.01CH37295).1*, pp. 508-512. Honolulu: Marine Technol. Soc. doi:10.1109/oceans.2001.968774

Freitag, L., Johnson, M., Singh, S., Grund, M., Kampf, S., Sherry, J., & Camilli, R. (2005). A long-range autonomous underwater vehicle for deep-ocean exploration at hydrothermal vents. *Journal of Field Robotics*, 22, 543–569.

Griffiths, G. (ed.). (2003). *Technology and Applications of Autonomous Underwater Vehicles*. Taylor & Francis. doi:10.1201/9780203023249

Hasan, H. (2017). *Underwater Robotics: Science, Design and Fabrication*. UTM Press: Malaysia.

Hegrenæs, Ø., Ramstad, A., Pedersen, T., & Velasco, D. (2016). Validation of a New Generation DVL for Underwater Vehicle Navigation. *IEEE/OES Autonomous Underwater Vehicles (AUV)* (pp. 342-348). Tokyo: IEEE. doi:10.1109/AUV.2016.7778694

Heshmat, M., Saoud, L. S., Abujabal, M., Sultan, A., Elmezain, M., Seneviratne, L., & Hussain, I. (2025). Underwater SLAM Meets Deep Learning: Challenges, Multi-Sensor Integration, and Future Directions. *Sensors*, 25(11). doi:https://doi.org/10.3390/s25113258

Hwangbo, S.-H., Jeon, J.-H., & Park, S.-J. (2013). Wireless Underwater Monitoring Systems Based on Energy Harvestings. *Sensors & Transducers, 18,* 113-119. Retrieved from https://www.researchgate.net/publication/290078574_Wireless_Underwater_Monitoring_Systems_Based_on_Energy_Harvestings/

Kilfoyle, D.B., & Baggeroer, A.B. (2000). The State of the Art in Underwater Acoustic Communication. *IEEE Journal of Oceanic Engineering.*

Liu, J., Yu, F., He, B., & Soares, C.G. (2024, April). A review of underwater docking and charging technology for autonomous vehicles. *Ocean Engineering, 297*(1). Retrieved May 2025, from https://www.sciencedirect.com/science/article/abs/pii/S0029801824004918

Martin, B., Tarraf, D. C., Whitmore, T. C., DeWeese, J., Kenney, C., Schmid, J., & DeLuca, P. (2019). *Advancing Autonomous Systems: An Analysis of Current and Future Technology for Unmanned Maritime Vehicles.* California: RAND Corporation. doi:10.7249/RR2664

Mohsan, S.A., Khan, M.A., Mazinani, A., Alsharif, M.H., & Cho, H.-S. (2022, September). Enabling Underwater Wireless Power Transfer towards Sixth Generation (6G) Wireless Networks: Opportunities, Recent Advances, and Technical Challenges. *Journal of Marine Science and Engineering, 10*(9). doi:https://doi.org/10.3390/jmse10091282

Moore, S., Bohm, H., & Jensen, V. (2010). *Underwater Robotics: Science, Design and Fabrication.* Monterey, CA: Marine Advanced Technology Education Center.

Multibeam Sonar. (n.d.). Retrieved January 2025, from NOAA Ocean Exploration: https://oceanexplorer.noaa.gov/technology/sonar/multibeam.html

Nkenyereye, L., Nkenyereye, L., & Ndibanje, B. (2024). Internet of Underwater Things: A Survey on Simulation Tools and 5G-Based Underwater Networks. *Electronics (*Special Issue: *Artificial Intelligence Empowered Internet of Things),* 13(3). doi:https://doi.org/10.3390/electronics13030474

Qiu, T., Zhao, Z., Zhang, T., Chen, C., & Chen, C. (2020). Underwater Internet of Things in Smart Ocean: System Architecture and Open Issues. *IEEE Transactions on Industrial Informatics,* 16, 4297–4307.

REMUS AUV. (n.d.). Retrieved May 2025, from Woods Hole Oceanographic Institution: https://www.whoi.edu/what-we-do/explore/underwater-vehicles/auvs/remus/

REMUS UUVs. (n.d.). Retrieved June 2025, from HHI: https://hii.com/what-we-do/capabilities/unmanned-systems/remus-uuvs/

Rice, J.A., & Frater, N. (2011). Optical Underwater Wireless Communications. *IEEE Journal of Oceanic Engineering.*

Rice, J.A., & Green, D. (2008). Underwater Acoustic Communications and Networks for the US Navy's Seaweb Program. *Sensor Technologies and Applications SENSORCOMM '08.* San Diego. doi:10.1109/SENSORCOMM.2008.137

Roberts, G.N., & Sutton, R. (eds.). (2013). *Further Advances in Unmanned Marine Vehicles.* The Institution of Engineering and Technology. doi:10.1049/PBCE002E

Rutherford, K.T. (2008). *Autonomous Underwater Vehicle Design Considering Energy Source Selection and Hydrodynamics.* University of Southampton.

Sentry. (n.d.). Retrieved June 2025 from Woods Hole Oceanographic Institution: https://www.whoi.edu/what-we-do/explore/underwater-vehicles/auvs/sentry/

Side Scan Sonar. (n.d.). Retrieved January 2025 from NOAA Ocean Exploration: https://

oceanexplorer.noaa.gov/technology/sonar/side-scan.html

Stojanovic, M. (2007). Underwater Acoustic Communication: Design Considerations on the Physical Layer. *IEEE Communications Magazine.*

Sub-Bottom Profiler. (2023, June 26). Retrieved from NOAA Ocean Exploration: https://oceanexplorer.noaa.gov/technology/sub-bottom-profiler/sub-bottom-profiler.html

Synthetic Aperture Sonar. (n.d.). Retrieved January 2025 from NOAA Ocean Exploration: https://oceanexplorer.noaa.gov/technology/sonar/sas.html

Teixeira, F.C., & Pascoal, A.M. (2007). Geophysical Navigation of Autonomous Underwater Vehicles. *IFAC Proceedings,* 40(17), 117-122. doi:https://doi.org/10.3182/20070919-3-HR-3904.00022

Tiano, A., & Ferri, G. (2018). Underwater vehicle navigation using SINS, DVL and LBL with intelligent sensor data fusion. *Journal of Intelligent & Robotic Systems,* 91, 263–278.

Tijjani, A. S., & Chemori, A. (2021). *Underwater Vehicles: Design and Applications.* Nova Science Publishers, Inc.

Wadoo, S., & Kachroo, P. (2011). *Autonomous Underwater Vehicles: Modelling, Control, Design, and Simulation.* CRC Press, Taylor & Francis Group.

Yuan, X., Martínez, J.-F., Pons, J.S., & Eckert, M. (2017, May). AEKF-SLAM: A new algorithm for robotic underwater navigation. *Sensors.* doi:doi: 10.3390/s17051174

Zhang, S., Zhao, S., An, D., Liu, J., Wang, H., Feng, Y., & Zhao, R. (2022, November). Visual SLAM for underwater vehicles: A survey. *Computer Science Review,* 46. doi:https://doi.org/10.1016/j.cosrev.2022.100510

Zhou, C. (2021). *Design of Autonomous Underwater Vehicles's Antenna System.* Stockholm: KTH Royal Institute of Technology. Retrieved from https://www.diva-portal.org/smash/get/diva2:1639312/FULLTEXT01.pdf

Zuluaga, C., Aristizábal, L., Rúa, S., Franco, D., Osorio, D., & Vásquez, R. (2022, March 25). Development of a Modular Software Architecture for Underwater Vehicles Using Systems Engineering. *Journal of Marine Science and Engineering,* 10, 464. doi:10.3390/jmse10040464

9

Autonomy, Control and Human Interaction

The ability of AUVs to operate without human intervention is at the core of their capabilities. Autonomy in underwater vehicles can be viewed as a spectrum, ranging from fully tele-operated systems (i.e., no onboard decision-making, such as ROVs) to fully autonomous vehicles that can perceive, decide, and act in dynamic environments. On the other hand, with increased operational complexity and extended mission durations, human-machine interaction (HMI) has also emerged as a pivotal aspect of AUV design.

Aspects of hardware and software for AUVs were introduced as components of vehicle design in chapter 4. In this chapter, we examine in greater detail, and for UUVs in general, various features of autonomy, hardware, software, control architecture and considerations for their interfaces with human operators.

The first section of this chapter introduces levels of autonomy. The next section outlines the system architecture in AUV design, covering the typical layers, modular and interoperable features, and protocols used across different layers. This is followed by a discussion on options for control architecture.

We touch upon various guidance and control algorithms for AUV control, introducing 'high-level' and 'low-level' algorithms and control approaches. These aspects ultimately determine their ability to execute missions in spite of the uncertainties of the real world.

The next two sections of this chapter discuss various aspects of human-AUV interaction and monitoring. We discuss design considerations for user interfaces (mission control). Data management solutions and architectures are discussed to handle the voluminous data generated from AUV operations.

At the end of the chapter we touch upon integration and system-level considerations, followed by future trends and research directions.

Levels of Autonomy

A widely accepted framework[1] categorises autonomy into levels listed below.

- **Level 0–Remotely Operated:** The vehicle is controlled entirely by a human operator via a tether or acoustic link. The onboard systems provide only basic stabilization and safety features. *Example:* Traditional ROVs used in deep-sea salvage or inspection.

- **Level 1–Basic Autonomy (Assisted Operation):** The vehicle is capable of executing simple, pre-programmed tasks (e.g., depth maintenance, straight-line cruising) and can react to basic environmental inputs, but still relies on human supervision for mission planning and decision-making. *Example:* Early generation AUVs with fixed mission scripts and limited sensor integration.

- **Level 2–Semi-Autonomous:** The AUV can adapt to environmental disturbances and perform dynamic adjustments using onboard sensor feedback. It is capable of executing mission segments with autonomy but still requires periodic operator intervention (e.g., for re-planning or emergency recovery). *Example:* Many modern glider-based platforms and workhorse AUVs like the REMUS series that perform survey missions with onboard obstacle avoidance and adaptive control.

- **Level 3–Fully Autonomous:** The vehicle is capable of high-level decision-making, including real-time mission re-planning, obstacle detection and avoidance, and execution of complex tasks without operator input. It integrates advanced sensor fusion, machine learning, and robust control algorithms to operate safely in uncertain environments. *Example:* Next-generation AUVs incorporating advanced AI for autonomous mapping, such as some prototypes developed for military and deep-sea exploration.

---

[1]　Huang, 2008

System Architecture

Rather than looking at UUVs as a collection of physical components (equipment, sensors, hardware, software), a more nuanced perspective looks at them as a system, with conceptual 'layers' based on functionality. Such a perspective is widely adopted for describing networks, emphasising the need to connect, exchange information and respond to stimuli.

The functionality of any UUV depends critically on a well-structured control architecture, which governs the interaction between software, hardware, and communication subsystems. This architecture determines how commands are executed, how sensor data is processed, how navigation and autonomy are achieved, and how modules interoperate—particularly in distributed and modular AUV platforms.[2]

Typical Layers of System Architecture

UUVs (AUVs and ROVs) can be considered as made up of distinct conceptual layers for hardware, software, and control. These layers (introduced for AUV design in chapter 4) work together to enable navigation, data acquisition, and mission execution in underwater environments.

Hardware Layer

For subsea systems (UUVs), this 'layer' represents the physical vehicle itself, with its propulsion system (thrusters, propellers, or buoyancy engines), sensors (pressure, depth, temperature, etc.), and onboard computers.

In the case of ROVs, we have systems outside the water (topside), and the connection to the ROV (umbilical). The topside systems (for ROVs) consist of the components on the surface vessel, such as the control console, power supply, and communication equipment. The umbilical cable is the physical link that transmits/receives data and transmits power from the surface to the underwater vehicle.

Software Layer

- **Operating System**: A real-time operating system manages the vehicle's resources and provides a platform for other software components.
- **Middleware**: This layer facilitates communication between different software modules and hardware components.

---

[2]　Antonelli, 2018

- **Mission-Specific Software:** This includes modules for navigation, path planning, obstacle avoidance, data acquisition, and communication protocols.
- **Sensor Data Processing:** Software algorithms process raw sensor data to provide usable information for navigation and other tasks.
- **Control Algorithms:** These algorithms translate desired vehicle movements into commands for the actuators.

Control Layer

- **Guidance System:** This layer determines the vehicle's overall path and objectives often involving path planning and behaviour selection.
- **Control System:** This layer implements the control algorithms to steer the vehicle, maintain depth, and execute manoeuvres.
- **Fault Detection:** The system monitors for errors and implements contingency plans to ensure mission safety.

Task Layer

- **Perception:** This involves using sensors to understand the environment and detect obstacles or features of interest.
- **Communication:** This handles data transmission between the vehicle and the surface (ROV) or other vehicles (AUV).
- **Motion Control:** This layer translates high-level commands into specific movements for the vehicle's actuators.
- **Fault Treatment:** This layer handles any errors or malfunctions that may occur during the mission.

Alternative Representations

AUV control architecture can be represented in terms of hierarchical layers in more than one way.[3] Another variant of the system architecture layers was introduced in chapter 4, and yet another variant is given below. Here, all layers refer to software or communication protocols for different aspects of AUV operation.[4]

---

[3] Caccia, 2006

[4] Yoerger, Jakuba, Bradley, & Bingham, 2007

- **Mission Layer**
  - o Responsible for high-level mission planning and decision-making.
  - o Implements logic for mission execution, goal prioritization, and contingency handling.
  - o May use behaviour trees, finite-state machines, or AI-based planners.

- **Task Execution Layer**
  - o Decomposes mission goals into executable tasks (e.g., waypoint following, data collection).
  - o Coordinates sensor payloads, control routines, and subsystems in response to mission needs.

- **Vehicle Control Layer**
  - o Executes control laws (e.g., PID, adaptive control) for guidance, navigation, and control (GNC).
  - o Interfaces with the vehicle's actuation systems (thrusters, fins).

- **Device and Hardware Abstraction Layer**
  - o Manages direct control over hardware components such as sensors, communication modules, and actuators.
  - o Provides hardware abstraction to ensure modularity and re-usability.

Modular and Interoperable Architectures

As AUV platforms become more modular and mission-flexible, there is a growing need for standardised communication and control interfaces. This ensures that new subsystems such as sonar modules, autonomy engines, or navigation sensors can be integrated with minimal effort. This is where 'middleware' and interoperability standards come into play.

Middleware

A robotics 'middleware' can essentially be thought as a software layer, composed of several modular packages, which has the task of collecting and harmonizing the information coming from the on-board sensors, making them available to the various processing nodes. These nodes must in turn process these data according to their specific function (tracking, communications, etc.) and pass the outputs to other nodes or directly to sensors and/or actuators of the robotic system. At the end of the various elaborations, the middleware will be responsible for the passage of the system in another state, e.g., the execution of a specific action.[5]

Some significant and common standards for control architecture are listed below:

- MOOS-IvP (Mission Oriented Operating Suite—Interval Programming)
- ROS (Robot Operating System)
- JAUS (Joint Architecture for Unmanned Systems)
- DDS (Data Distribution Service)

The current de facto standard middleware for UUVs include MOOS and ROS, which are introduced here. Both MOOS and ROS are publish-and-subscribe systems, which provide the communication of arbitrary data throughout a network.[6]

MOOS-IvP

MOOS is a lightweight middleware framework developed by the Oxford Mobile Robotics Group, designed specifically for marine robotics, maintained by MIT. It is optimized for mission control, real-time coordination, and long-duration, autonomous operations. In order to complete mission objectives, a robotic system also requires a deliberative component in addition to its reactive aspects (e.g., avoiding obstacles). This is the objective of the MOOS process 'IvP-Helm', which acts as an autonomous decision-making engine to execute in the 'backseat' of the robotic platform.

MOOS-IvP stands for 'Mission Oriented Operating Suite–Interval Programming'. MOOS-IvP includes IvP Helm, a behaviour-based decision engine used for high-level autonomy. MOOS-IvP is ideal for deploying in bandwidth-limited marine environments, or for behaviour-based mission execution.[7]

---

5   Costanzi, Fenucci, & Manzari, 2020
6   Benjamin, Schmidt, Newman, & Leonard, 2010
7   MOOS-IvP; https://oceanai.mit.edu/moos-ivp/pmwiki/pmwiki.php?n=Main.HomePage.

ROS

ROS (Robot Operating System) is a flexible, modular middleware framework for developing robot software. It is not an actual operating system, but a collection of tools, libraries, and conventions that simplify the task of building complex and scalable robot systems. Originally developed for ground robots, ROS is now widely used in marine robotics, especially for rapid prototyping, data handling, and integration of heterogeneous subsystems.[8]

In ROS, data are transferred using peer-to-peer communications (on the other hand, all data within a MOOS system are transmitted through the MOOSDB). The basic unit of interaction in ROS is a 'message', which is typically exchanged on a 'topic': from a node's perspective, messages are published synchronously and read from a subscription asynchronously. In ROS, there are two extensions to the typical publish/subscribe mechanism: 'services' and 'actions'. A 'service' simply takes an input and returns an output, as a typical service does in computer science literature. An 'action' keeps an internal state on a longer time scale because it is called with a goal, emits feedback during the action, and finally returns a single result at the end of the action. An important advantage of ROS over MOOS is its capability of handling very large datasets, e.g., live high-resolution video and multi-dimensional point clouds from LIDAR-like devices.

Tools available in ROS include visualization (RViz), simulation (Gazebo), logging, diagnostics, and bagfile recording. It has an extensive library including SLAM, computer vision, path planning, and control algorithms. Its subsequent version, ROS 2, adds support for real-time, DDS-based communication, and better multi-robot systems.

DDS

DDS (Data Distribution Service) is a standardized middleware protocol and API (defined by the OMG–Object Management Group) for real-time, scalable, high-performance communication between distributed systems.

Unlike ROS and MOOS, which are frameworks with many tools and concepts, DDS is purely a communication layer like a 'transport backbone', designed to: Deliver real-time, low-latency, and reliable data exchange; Enable decentralized, peer-to-peer communication; and Support quality of service

---

[8]   DeMarco, West, & Collins, 2011

(QoS) controls (latency, reliability, history, deadlines, etc.). ROS2 is built natively on DDS (while ROS 1 does not use DDS by default). MOOS does not use DDS, and uses its own internal centralised bus (MOOSDB), where all components publish/subscribe to named variables.

JAUS in AUV Control Architectures

JAUS (Joint Architecture for Unmanned Systems) is a specification introduced by the US DoD to provide a basis for interoperability between unmanned systems.[9] Developed under SAE AS-4 standards, JAUS is a message-based, platform-independent standard for the command and control of unmanned systems, particularly military AUVs. It defines a set of standard interfaces, messages, and component behaviour to ensure interoperability across heterogeneous unmanned systems: land, air, or under water.

A set of standard application layer interfaces called 'JAUS Mobility Services' have been defined. JAUS Services provide the means for software entities in an unmanned system (or system of unmanned systems) to communicate and coordinate their activities. The Mobility Services represent vehicle platform-independent capabilities commonly found across all domains and types of unmanned systems (referred to as UxVs). At present, over 15 services are defined in this document, many of which have been updated to support Unmanned Underwater Vehicles (UUVs).[10]

Based on JAUS standards, software development kits (e.g., C++ based middleware toolkit) are available commercially.[11] Such middleware can be used in an AUV so as to standardise its software in order to make it interoperable with other SAE JAUS-based systems. Open source options are also available.[12]

In the context of AUVs, JAUS has the following features:

- **Interoperability and Modularity:** JAUS promotes a component-based architecture, where subsystems (e.g., navigation, propulsion, sensor management) are represented as software components with well-defined interfaces. This allows AUV developers to integrate

---

[9] Whitsitt & Sprinkle, 2011

[10] JAUS Mobility Service Set AS6009A, 2023

[11] https://openjaus.com/sdk/

[12] https://github.com/jaustoolset/jaustoolset

components from multiple vendors, replace or upgrade subsystems without affecting the rest of the system, and simplify validation and testing via defined interface behaviour.[13]

- **Distributed Communication**: JAUS supports peer-to-peer messaging, enabling distributed control. In an AUV, this allows the autonomy engine to directly communicate with the propulsion system, mission management systems to dynamically reconfigure payloads, and sensor fusion nodes to interact seamlessly with localization systems.

- **Multi-Vehicle Coordination**: When operating as part of a fleet (e.g., swarm missions), JAUS supports coordinated behaviour by standardizing messages for position updates, task delegation, health monitoring and status reporting.

- **Safety and Diagnostics**: JAUS includes message sets for health monitoring, fault reporting, and emergency control. These are essential for underwater vehicles that may be deployed for extended periods in unstructured environments.

Comparison of Middleware/Frameworks

The development of MOOS-IvP in the research community is continuous and ongoing but its usage in industry is extremely limited. This may be a consequence of the lack of insurance on backward compatibility, which is based on community agreement and not on enforced standards. While MOOS has historically been popular within the underwater robotics community, ROS is now by far more pervasive in a multi-domain context (ground, sea, air). The main reasons are that ROS is reconfigurable and relatively easy to use. There are bindings for both C++ and Python, and it is regularly updated (at least yearly). There are no rigid standards guiding development, but it is so widely used that there is extreme prejudice against breaking backward compatibility.[14] The main features of MOOS-IvP and ROS are compared in Table 9-1.

---

[13]   Costanzi, Fenucci, & Manzari, 2020

[14]   Costanzi, Fenucci, & Manzari, 2020

Table 9-1. Comparison of ROS and MOOS Middleware

Feature	*ROS*	*MOOS (MOOS-IvP)*
Design Origin	Ground/mobile robotics (general)	Marine robotics, mission control
Communication	Pub-Sub via Topics, Services	Central database with variable updates
Autonomy Engine	Custom (or via Behaviour Trees/AI)	IvP Helm (multi-objective behaviour engine)
Real-Time Support	ROS 1: No; ROS 2: Yes (via DDS)	Yes (simple, deterministic message system)
Ease of Use	Rich libraries, tools, rapid dev	Lightweight, but requires C++ expertise
Field Proven In	Prototypes, commercial AUVs, drones	Long-range AUVs, WHOI missions, swarms

The main features of the protocols, software framework, middleware and interoperability standards discussed above (MOOS, ROS, DDS and JAUS, respectively), are compared in Table 9-2 for further clarity about their role, scope and usage.

Other Considerations

AUVs may use either 'hierarchical' or 'behaviour-based' control architectures. ROVs typically rely on a hierarchical structure, since they operate (almost) always under human control. Other than hierarchical, control architectures could be characterised as 'heterarchical', 'subsumption', or 'hybrid'.

Communication networks are a major feature of any system architecture. Robust communication is vital for ROVs (through the umbilical) and AUVs (using acoustic modems or other wireless technologies).

Sensor fusion has been mentioned in the earlier chapter on navigation. AUVs often fuse data from multiple sensors (e.g., INS, GPS, sonar) to improve navigation accuracy.

Protocol Stack

A protocol stack is a layered model showing how data flows from a high-level application (like mission control software) down to the physical communication hardware (e.g., an acoustic modem), and vice versa.

Table 9-2. Comparison of popular protocols and frameworks

Feature	*DDS*	*ROS (1 & 2)*	*MOOS*	*JAUS*
Description	Middleware protocol	Robot software framework	Lightweight middleware + autonomy framework	Interoperability and messaging standard
Design Origin	Real-time distributed systems (OMG)	Ground/mobile robotics (Willow Garage, OSRF)	Marine robotics (Oxford MRG, WHOI)	DoD/SAE standard for unmanned systems
Communication Model	Peer-to-peer, publish/subscribe, QoS-controlled	ROS 1: Centralized pub-sub ROS 2: Peer-to-peer via DDS	Centralized variable store (MOOSDB)	Peer-to-peer message passing (strict message formats)
Standardisation	Yes (OMG DDS)	ROS 1: Community-based ROS 2: OMG DDS-compliant	No formal standard (de facto in marine research)	Yes (SAE AS-4 series standards)
Real-Time Support	Yes, deterministic with QoS	ROS 1: No ROS 2: Yes (through DDS)	Yes (simple, predictable scheduling)	Yes, deterministic messaging
Scope	Transport layer only	Full robotics stack (control, planning, perception)	Mission-level control, autonomy behaviours	Messaging structure, service definitions for unmanned systems
Modularity/ Inter-operability	Moderate (depends on DDS vendor)	ROS 2: High ROS 1: Moderate	Moderate (tight coupling to MOOSDB)	High (focus on cross-component interfaces)
Autonomy Support	Not included (use with higher layers)	Yes (via planners, behaviour trees, AI integration)	Yes (IvP Helm behaviour engine)	Limited (focus on control, not behaviour logic)
Ease of Use	Complex to configure	ROS 1: Easy ROS 2: Moderate (more robust)	Moderate (C++ based, compact APIs)	Steep learning curve, rigid specification
Field Proven In	Defence systems, ROS 2 back ends, DDS-native systems	Prototypes, AUVs, drones, industrial robotics	WHOI AUVs (REMUS), academic fleets	US DoD AUV/UUV programs, Kingfish, Knifefish, etc.
Primary Strengths	Deterministic, decentralized, scalable	Rich ecosystem, modularity, perception support	Lightweight, reliable, mission-focused	Standardized, interoperable control across vendors
Weaknesses	No autonomy tools, requires DDS vendor/ runtime	ROS 1: not real-time; ROS 2: configuration is complex	Less scalable, no advanced perception support	Rigid message format, limited dynamic behaviour execution

Protocol layer placement in an AUV/ROV stack is shown below, with examples:

```
High-level Autonomy & Control ]
    |____ JAUS, ROS, MOOS-IvP, DDS, IMC
[ Transport & Messaging ]
    |____ DDS-RTPS, SOME/IP, TCP/UDP, NMEA
[ Physical Communication Interfaces ]
    |____ RS-232, RS-485, CAN, I²C, SPI
[ Specialized Underwater Comms ]
    |____ JANUS (acoustic), WHOI Micromodem, BlueComm (optical)
```

A typical protocol stack in the context of AUV communication is shown below.

High-Level Software

$\updownarrow$

Middleware/Control Protocols (e.g., MOOS, ROS, DDS)

$\updownarrow$

Transport Layer (e.g., TCP/UDP)

$\updownarrow$

Acoustic Protocol Layer (e.g., S2C LLP, WHOI micro-modem packet format)

$\updownarrow$

Modem Hardware Interface (RS-232, Ethernet, etc.)

$\updownarrow$

Physical Layer (Sound propagation in water)

Control and communication protocols used in AUVs/ROVs extend across different layers of the system architecture, as shown in Table 9-3. It is essential to know which protocol(s) is/are supported by particular equipment, in order to determine whether (or how) it can be integrated into the UUV being designed or developed.

Table 9-3. Protocols used at different layers of system architecture

Layer	Examples	Function
Application Layer (High-level)	JAUS, ROS, DDS, MOOS-IvP	Autonomy, control logic, inter-module messaging
Transport/Messaging Layer	TCP/UDP, SOME/IP*, DDS-RTPS	Message encoding, delivery across subsystems
Hardware/Bus Communication Layer (Low-level)	SPI, I²C, CAN, RS-232	Communication between sensors, controllers, and low-level boards

*Used in vehicles, not in UUVs

Control Strategies

Control strategies refer to algorithms or techniques that drive decision-making and actuation in autonomous systems. These strategies transform intent (goals) into action (movement or behaviour) while dealing with constraints, uncertainty, and changing environments.

Control strategies are broadly classified into:

- High-Level Control Strategies
  - o Focus on mission planning, decision-making, behaviour arbitration, and task sequencing
  - o Typically operate at the task layer
  - o Tend to be deliberative or reactive logic governing what to do next
- Low-Level Control Strategies
  - o Focus on motion control, stabilization, tracking, and actuator coordination
  - o Typically operate at the control layer
  - o Tend to be mathematical or feedback-based algorithms for how to move or behave

AUV control systems combine both low-level and high-level control strategies. ROS (1 and 2) and MOOS IvP both support high-level control strategies. JAUS provides a standardised interface for control commands, but only supports messaging for high/low-level control. DDS does not natively support high/low level control, but supports comms for any custom control architecture.

We now look at high-level, followed by low-level, control strategies for AUVs.

High-Level Guidance and Decision-Making

High-level guidance algorithms plan missions and generate trajectories for the low-level controllers. They are located at the *Task* layer. They decide *what* to do, *when* to do it, and *how* to recover from faults. They operate at timescales of seconds to minutes. Examples are as follows:

- **Waypoint Navigation:** Involves planning a path that connects predefined waypoints. This is common in survey missions.

- **Behaviour-Based Control:** Decomposes complex missions into simpler behaviours (e.g., 'obstacle avoidance,' 'data collection,' 'energy conservation') that are triggered based on environmental cues.

- **Fuzzy Logic Control:** Uses linguistic rules to handle uncertainty and nonlinear dynamics in the decision-making process.

- **Artificial Intelligence (AI) and Machine Learning (ML):** Emerging techniques that allow for autonomous adaptation and mission re-planning based on sensor data, historical performance, and environmental changes.

A comparison of salient high-level control strategies is shown in Table 9-4.

Table 9-4. Summary of High-level control strategies for AUVs

Control Approach	Type	Key Features	Typical Applications (for UUVs)
Finite State Machines (FSM)	Rule-Based/ Discrete Logic	Simple, deterministic transitions between states	Mission sequencing (e.g., descend → survey → surface)
Behaviour Trees	Modular Reactive Planning	Hierarchical, modular, readable, interruptible	Flexible mission planning, adaptable behaviours
IvP Helm (MOOS-IvP)	Multi-Objective Behaviour Arbitration	Optimizes weighted behaviours at runtime, supports dynamic priorities	WHOI-style autonomy: line following + collision avoidance + search
Hierarchical Task Networks (HTN)	Symbolic Planning	Decomposes tasks into subtasks recursively, logic-driven	Complex mission plans, automated planning in structured scenarios
Deliberative Planning (e.g., A, D, RRT)**	Search-Based Planning	Graph-based planning, maps environment, generates optimal paths	Pre-mission path planning, obstacle-aware navigation
Reactive Planning (e.g., VFH, DWA)	Real-Time Reactive Control	Uses sensor data for local obstacle avoidance in real time	Terrain following, dynamic obstacle avoidance
AI/ML-Based Planning	Learning-Based Autonomy	Learns mission policies or decision models from data or simulations	Adaptive re-tasking, mission optimization in unknown conditions
Swarm/Fleet Coordination	Distributed Task Allocation	Assigns tasks across multiple vehicles, often using auctions or consensus	Multi-UUV mapping, coverage, cooperative search
Goal-Oriented Action Planning (GOAP)	Goal-Driven Symbolic Planning	Flexible plans generated from goals and current world state	Dynamic goal execution and re-planning under environmental changes

Finite state machines (FSMs) and IvP Helm are mature and widely deployed in fielded UUVs. Behaviour Trees offer a more maintainable and scalable alternative to FSMs. AI/ML-based planning is still emerging but increasingly valuable in adaptive autonomy. Swarm coordination strategies are vital for next-generation UUV systems to operate in fleets or MCM (mine countermeasure) missions.

Low-Level Control Strategies

These algorithms operate in the inner control loops, maintaining vehicle stability and following short-term trajectories. They operate at timescales of milliseconds to seconds. They are located at the control layer. Their functions include stabilising depth, heading, and speed.[15] Examples of such algorithms are mentioned below:

- **PID Controllers:** Widely used due to their simplicity and effectiveness. They provide corrective feedback for errors in depth, heading, and speed.

- **Sliding Mode Control (SMC):** Offers robustness in the presence of model uncertainties and external disturbances, making it well-suited for harsh underwater environments.

- **Model Predictive Control (MPC):** Uses an explicit model of the vehicle dynamics to optimize control actions over a prediction horizon, thereby handling multivariable constraints and nonlinearities.

- **Adaptive Control:** Adjusts controller parameters in real time to cope with changing dynamics, such as variations in drag or payload.

- **Behaviour-based Control:** Often acts as a bridge between high-level decisions and low-level actuation, especially in reactive systems.

- **AI/ML-based Methods** are becoming popular for adaptive or complex environments, though their trustworthiness and certifiability are active research areas in safety-critical UUV applications.

Table 9-5 summarises common low-level control strategies used in AUVs, along with key features and typical applications.

---

[15] Fossen, 1991

Table 9-5. Low-level Control Strategies for AUVs

Control Approach	Type	Key Features	Typical Applications (for UUVs)
PID Control	Feedback (Linear)	Simple, widely used, good performance in stable conditions	Depth, heading, speed control
Back stepping Control	Nonlinear Control	Lyapunov-stable, handles system nonlinearities	Fin-actuated control, roll/pitch stabilization
Sliding Mode Control	Robust Control	Strong robustness to disturbances, discontinuous control	Harsh environments, actuator fault tolerance
Linear Quadratic Regulator (LQR)	Optimal Control	Optimal trade-off between control effort and error	Smooth path tracking under linear approximations
Adaptive Control	Online Parameter Estimation	Adjusts to changing dynamics or uncertainties	Payload changes, long missions with evolving drag/mass
Model Predictive Control (MPC)	Predictive Optimal Control	Future-aware, constraint handling, high precision	Obstacle avoidance, advanced manoeuvring in cluttered environments
Fuzzy Logic Control	Rule-Based/ Heuristic	Handles vague or imprecise models, intuitive tuning	Buoyancy control, variable ballast systems
H∞ Control	Robust Optimal Control	Worst-case optimization under uncertainty	Military or critical missions with unknown disturbances
Feedback Linearization	Nonlinear Control	Cancels out nonlinear terms to linearise dynamics	High-performance control with accurate models
Behaviour-Based Control	Hybrid/Reactive Architecture	Uses modular behaviours (e.g., avoid, follow, search), often with arbitration	Terrain following, object tracking, reactive obstacle avoidance
AI/ML-Based Control	Data-Driven/ Learning-Based	Learns control policies from data or simulation (e.g., RL, imitation learning)	Adaptive navigation, unstructured environments, mission re-planning

Guidance, Navigation and Control

In chapter 4 (AUV design), we introduced the concepts of guidance, navigation and control (GNC) in the context of hydrodynamics and the vehicle's mathematical model of motion. We now revisit GNC to see their relation with middleware frameworks and control strategies of the AUV.

Definitions

- Guidance: Determines where the vehicle should go, e.g., path planning, waypoint navigation.
- Navigation: Estimates position from sensors (IMU, DVL, sonar)

- Control: Determines how the vehicle moves, e.g., depth control, heading stabilization, thruster commands.

Together, these form the Guidance, Navigation, and Control (GNC) subsystem.[16]

In a layered AUV control architecture, guidance and control algorithms typically operate at the Control Layer, which is situated between high-level mission logic and low-level device drivers. The GNC translates *mission-level commands* into *actuator-level signals*.

Relation to Middleware Frameworks

The middleware frameworks (ROS, MOOS, JAUS, and DDS) that we discussed connect the guidance/control algorithms with sensor data inputs (e.g., position, velocity, orientation); mission/task commands (e.g., new waypoint, change depth); and actuator interfaces (e.g., thruster set points). The summary of roles of GNC and middleware given in Table 9-6 will help in clarifying these concepts.

Table 9-6. Summary of GNC module roles and their interaction

Component	*Role in GNC*	*Interaction via Middleware*
Guidance Module	Plans path, selects desired heading/speed	Receives waypoints, sends goals to control module
Control Module	Executes motion using control laws (e.g., PID)	Converts goals to actuator signals
Navigation Module	Estimates position from sensors (IMU, DVL, sonar)	Publishes pose for use by guidance/control
Middleware	Connects GNC modules to sensors, actuators, and autonomy engine	ROS: topics/services MOOS: MOOSDB vars JAUS: standard messages

Thus, middleware enables the GNC system to be modular and distributed. Control architecture organises GNC within a layered structure, maintaining separation of concerns. JAUS, ROS, MOOS, and DDS all serve to support the communication and orchestration of GNC algorithms, but at different layers and with different philosophies.

---

[16] De Barros, Pascoal, & De Sa, 2004

Linkage with Control Strategies

Control strategies are central to GNC, but also extend into autonomy, coordination, and mission execution. Low-level control stabilizes and moves the vehicle. High-level control makes decisions and governs complex behaviours. Middleware like ROS, MOOS, and JAUS enables the communication, distribution, and modular deployment of these strategies. Control strategies are not limited to GNC, but GNC is the most mathematically intensive and tightly coupled domain for them.

Most low-level control strategies (discussed above) fall directly under GNC. However, high-level strategies often span areas other than GNC as well. A few examples are given in Table 9-7.

Table 9-7. Use of High-level Control Strategies for GNC and other areas

Domain	*Strategy Type*	*Examples*
Guidance	Low/High-level	Waypoint tracking, path planning, collision avoidance
Navigation	Support layer	Kalman filter, SLAM, dead reckoning
Control	Low-level	PID, LQR, back stepping, MPC, adaptive control
Autonomy & Mission Logic	High-level	Behaviour trees, state machines, IvP Helm
Multi-AUV Coordination	High-level	Swarm control, task allocation, leader-follower
Payload/ Sensor Control	High or low-level	Adaptive sampling, sonar control, power management

Human–AUV Interaction

Human operators interact with AUVs in a variety of ways depending on the level of autonomy, mission requirements, and operational environment. This section examines the modalities of interaction, the communication channels used, and the monitoring techniques that enable operators to maintain situational awareness while remotely supervising underwater missions.[17]

---

[17] Moore, Bohm, & Jensen, 2010

Modes of Interaction

Direct Tele-operation

In early AUV operations, human operators were required to control the vehicle directly. This mode, often referred to as tele-operation, involves continuous manual control of the AUV's motion and sensor activation via an interface (typically a joystick or keyboard input) and real-time data feedback from the vehicle. Tele-operation has several advantages, including:

- **Immediate Response:** Operators can instantly adjust the vehicle's trajectory or sensor settings based on visual feedback.

- **Fine Control:** Direct control allows for precise manoeuvring in cluttered or dynamic environments.

- **High Situational Awareness:** Continuous input and monitoring enhance the operator's connection to the vehicle's current state.

However, tele-operation is labour-intensive and places a high cognitive load on the operator, particularly during long-duration missions. Additionally, the limitations of underwater communication—such as limited bandwidth and intermittent connectivity—can compromise the effectiveness of direct tele-operation.

Supervisory Control and Shared Autonomy

As AUV technology matured, control strategies shifted towards supervisory control and shared autonomy models. In these roles, the AUV executes pre-programmed mission plans autonomously, while the human operator oversees the overall mission and intervenes only when necessary. This approach offers several benefits:

- **Reduced Workload:** Operators can monitor multiple vehicles simultaneously without needing to control every movement.

- **Increased Scalability:** A single operator can supervise a fleet of AUVs, which is particularly important in large-scale operations.

- **Energy Efficiency:** Autonomous operation allows the AUV to optimize its energy use without frequent human intervention.

In supervisory control, the human-machine interface (HMI) typically provides a high-level summary of each vehicle's status, including key performance indicators such as battery levels, position, sensor health, and error states.

Operators can then intervene—often through a graphical user interface (GUI) that allows them to issue commands (e.g., 'hold position' or 'return to base') if anomalies are detected.

Hybrid Interaction Models

Modern systems increasingly employ hybrid interaction models that blend direct control with supervisory oversight. These systems allow for different levels of control based on mission context. For instance, during critical phases of a mission (e.g., obstacle avoidance or emergency manoeuvres), the system may switch to a lower level of autonomy with direct operator input. Conversely, during routine survey operations, the system may rely on full autonomy with only periodic status updates from the vehicle.

Hybrid models require adaptable interfaces that can dynamically change the level of detail presented to the operator. For example, when the AUV is in safe cruising mode, the interface might display aggregated data (e.g., a heat map of the vehicle's path or a summary of sensor readings). However, if the system detects an anomaly, the interface could switch to a more detailed view, displaying live video feeds, sensor data streams, and diagnostic information.

Communication Channels

Operators interact with AUVs using a combination of communication channels:

- **Acoustic Communications:** Underwater, the primary mode of communication is acoustic. Acoustic modems can transmit telemetry data and command messages over distances of up to several kilometres, though at relatively low data rates. Acoustic communications are subject to environmental interference, multipath propagation, and latency.

- **RF and Satellite Links:** When AUVs surface or operate near a surface buoy, RF communications (including Wi-Fi or cellular networks) and satellite links may be used to transmit high-bandwidth data, such as video feeds or detailed sensor imagery.

- **Local Storage and Post-Mission Download:** In many cases, AUVs store large amounts of data on-board for later download. Operators then use post-mission analysis tools to review the data in greater detail.

In supervisory control scenarios, real-time data are often summarized and compressed to conserve bandwidth. For instance, rather than streaming full video feeds continuously, the system might transmit key frames or use event-driven reporting, where only anomalies or mission-critical events are reported in real time.

Monitoring Techniques

Effective monitoring of AUVs involves a combination of sensor data aggregation, data fusion, and visualization techniques:

- **Dashboard Displays:** Mission control centres typically feature dashboards that consolidate critical parameters (e.g., battery life, depth, position, sensor status) into an easily digestible format. Graphical elements such as gauges, charts, and status indicators allow operators to quickly assess vehicle health.

- **Map-Based Interfaces:** Many systems use digital maps to display the AUV's real-time position, planned mission route, and environmental data. Interactive map interfaces enable operators to zoom in on areas of interest, view historical tracks, and identify potential hazards.

- **Alert Systems:** Automated alerts (visual, auditory, or haptic) notify operators when certain parameters exceed predefined thresholds. For example, an alert may be generated if the AUV's battery level drops below a critical value, or if the vehicle detects unexpected obstacles.

- **Data Fusion and Analytics:** Advanced systems integrate data from multiple sensors to produce higher-level interpretations of the underwater environment. For example, combining sonar and video data can improve target detection and obstacle avoidance. Such fused data are then presented in a summarized form to the operator.

Together, these monitoring techniques allow human operators to maintain a high level of situational awareness even when controlling multiple AUVs, thus ensuring mission safety and success.

User Interfaces for Mission Control

Designing user interfaces for AUV mission control presents unique challenges due to the complex operational environment and the need to manage large amounts of heterogeneous data. This section outlines the key design principles and considerations that inform the development of effective mission control interfaces.

Design Principles

Situational Awareness

One of the most critical objectives of any mission control interface is to support operator situational awareness (SA). SA is defined as the perception of elements in the environment, comprehension of their meaning, and projection of their status in the near future. Effective interfaces should:

- **Display Critical Information Clearly:** Use well-organized dashboards and graphical displays that highlight essential parameters (e.g., vehicle position, energy levels, and sensor health).
- **Hierarchical Data Presentation:** Information is organized hierarchically, with high-priority data at the top and options to drill down into detailed data. This reduces clutter and allows the operator to focus on mission-critical information.
- **Employ Multimodal Feedback:** Combine visual, auditory, and, when possible, haptic feedback to reinforce the operator's understanding of the AUV's state.
- **Minimize Cognitive Load:** Present information in a simplified and aggregated form, using methods such as heat maps or trend lines to reduce the need for constant interpretation of raw data.

Research on human-swarm interaction has shown that using visual abstractions like heat maps significantly reduces the operator's cognitive burden when managing multiple vehicles.

Usability and Ergonomics

The effectiveness of an interface is strongly linked to its usability. Reducing the cognitive workload of the human operator is a primary design goal. Research in human-swarm and human-robot interaction emphasizes that user interfaces must be designed to support both high-level supervision and detailed control without overwhelming the operator. Future systems may incorporate artificial intelligence to predict operator workload and adjust interface complexity in real time. In mission-critical operations, the interface must be intuitive and easy to use, for which the salient features are listed below.

- **Consistency:** Interfaces should follow a consistent layout and design language to allow rapid familiarization.
- **Responsiveness:** Feedback should be immediate, with minimal delay between operator command and system response.
- **Adaptability:** Interfaces may need to adapt dynamically depending on the phase of the mission. For example, an interface might provide more detailed controls during critical manoeuvres and revert to summary views during routine operation.
- **User Customization:** Allowing operators to customize their displays (e.g., choosing which parameters to monitor, adjusting display scales) can greatly improve performance.
- **Error Prevention and Recovery:** The design should include mechanisms for error detection, clear notifications, and easy recovery options (e.g., one-click emergency stop or reset functions).

User studies in related domains (e.g., UAV swarm management) have shown that interfaces that allow a high degree of customization and adapt to user preferences result in lower cognitive load and fewer operational errors.

Information Prioritization

Given the volume of data generated by AUV sensors, prioritising the display of information is crucial. Not all data are equally important at every stage of a mission. Effective strategies include:

- **Layered Information Display:** Use hierarchical displays that show high-priority data at the top level and allow the operator to drill down for more detail as needed.
- **Alert and Notification Systems:** Highlight critical alerts using colour codes or flashing indicators to ensure the operator's attention is drawn to the most important issues.
- **Context-Aware Display:** The interface should dynamically adjust the information shown based on the current operational context. For example, during low-risk phases, the interface may display only summary statistics; during emergencies, detailed sensor readouts and control options should become available.

Transparency and Trust

Operators must trust the system to make autonomous decisions while still retaining ultimate control:

- **Data Transparency:** Display not only the decisions made by the autonomous systems but also the underlying sensor data and status information that led to these decisions.

- **Explainability:** When possible, provide brief explanations for automated actions so that operators can understand why certain commands were issued.

- **Intervention Options:** Ensure that operators have the ability to override automated decisions easily if they detect any discrepancies or anomalies.

Studies on human-machine interaction emphasize that interfaces that enhance transparency lead to increased trust and better performance in supervisory control scenarios.[18]

Safety and Reliability

Safety and reliability are paramount in AUV operations. Features towards these requirements are listed below.

- **Redundant Systems:** Many AUVs are designed with redundant sensors and communication channels to ensure continuous operation in the event of a component failure.

- **Error Detection and Recovery:** Automated error detection algorithms monitor the health of the system and trigger appropriate recovery procedures if anomalies are detected. In such cases, the interface notifies the operator and provides recommendations for corrective action.

- **Transparency:** The system should provide clear explanations for automated decisions to build operator trust. This includes displaying diagnostic information and the rationale behind specific actions taken by the AUV.

- **Interoperability Standards:** Adopting industry standards (e.g., JAUS, STANAG) enhances interoperability between systems, contributing

---

[18] Soorati, Clark, Ghofrani, Tarapore, & Ramchurn, 2021

to overall reliability and ensuring that operators can manage multiple systems with consistent interfaces.

Mission Control Interfaces

Technologies and tools used for mission control interfaces are discussed in this section.

Graphical User Interfaces (GUI)

Modern mission control centres use sophisticated GUIs to present real-time data, as listed below.

- **Dashboards:** Often include multiple panels such as status indicators, maps, video feeds, and sensor data graphs.

- **Touchscreen Interfaces:** Many systems incorporate touchscreen controls, enabling intuitive interactions such as drag-and-drop commands for setting waypoints.

- **Augmented Reality (AR) and Virtual Reality (VR):** These technologies are increasingly used to present 3D visualizations of the underwater environment, providing operators with a more immersive view and improved depth perception. For instance, AR overlays can help highlight potential obstacles or mission-critical areas.

Software Platforms and Middleware

Several software platforms have been developed to support mission control:

- **Ground Control:** An open-source ground control station originally designed for unmanned aerial vehicles (UAVs) but extended to support underwater vehicles. It integrates telemetry data, mission planning, and vehicle status information into one platform.

- **VideoRay Cockpit:** A proprietary interface used for remote operations of ROVs and AUVs that emphasizes ease of use and robust data integration.

- **Custom Middleware:** In many cases, specialized middleware is developed to fuse data from multiple sensors, perform on-board processing, and relay information in an energy-efficient manner. This middleware must be designed with interoperability and scalability in mind.

Human Factors and Ergonomic Considerations

Ergonomic design is critical and certain salient considerations are given here.

- **Display Layouts:** Well-organized and minimally cluttered layouts help reduce visual fatigue and enhance operator performance.

- **Font and Colour Choices:** High contrast, legible fonts, and carefully chosen colour schemes (e.g., red for alerts, green for normal status) ensure that operators can quickly interpret information even in high-stress situations.

- **Input Devices:** Operators often use specialized controllers or keyboards optimized for rapid command entry and error-free operation.[19]

Case Studies and Examples

Several operational systems demonstrate effective mission control interfaces:

- The mission control system for the MARIUS AUV provides a good example of a Petri net-based framework that allows for both high-level mission planning and low-level control. The system enables operators to monitor vehicle status and intervene when necessary, using a combination of graphical displays and textual command interfaces.[20]

- Recent innovations in AUV interfaces, such as those implemented on Kongsberg's HUGIN Endurance, highlight the shift towards adaptive interfaces that can switch between summary and detailed views based on the mission phase.

Integration and System-Level Considerations

The integration of human interface design, energy management, and data processing forms the backbone of effective AUV mission control systems. In this section, we outline how these components are combined into a cohesive architecture that supports mission-critical operations.

---

[19]   Abdullah, Chen, Blow, & Uthai, 2024
[20]   Oliveira, Pascoal, Silva, & Silvestre, 1998

Figure 9-1. Operator interfacing with UUV
(Source: NIO Goa)

Mission Control System Architecture

Modern mission control systems for AUVs typically include:

- **Operator Workstations:** These are equipped with adaptive GUIs that present real-time telemetry, sensor data, energy status, and diagnostic alerts. The design of these workstations is informed by human factors research to ensure minimal cognitive load.

- **On-Board Control Systems:** These include processors that perform sensor fusion, execute energy management algorithms, and implement adaptive communication protocols. On-board systems also handle local decision-making for tasks such as obstacle avoidance and mission adaptation.

- **Communication Middleware:** Middleware ensures seamless data exchange between the AUV and the operator. It supports multiple communication modes, automatically switching between low-power acoustic and high-bandwidth RF links as required.

- **Data Management Backends:** These systems handle both on-board storage and post-mission data transfer to Cloud or centralized repositories. They support advanced analytics and long-term data mining.

A robust system architecture must integrate these components seamlessly. For instance, if the AUV's energy management system detects low battery levels, it can trigger an adaptive communication mode that reduces data transmission frequency. Simultaneously, the operator workstation is updated to reflect the energy status, enabling the operator to decide whether or not to alter the mission plan.

Key Design Considerations

Human interaction with AUVs is a multifaceted challenge that encompasses interface design, energy management, and data processing. Although AUVs have reached high levels of autonomy, human operators remain essential for mission planning, real-time monitoring, and emergency interventions. Effective mission control interfaces must be designed to support situational awareness, reduce cognitive load, and facilitate rapid decision-making while balancing the constraints imposed by limited energy resources. Key design considerations include:

- The adoption of adaptive, context-aware interfaces that dynamically change based on mission phase and operator workload.
- The implementation of intelligent communication protocols that conserve energy while ensuring that critical data are transmitted promptly.
- The use of on-board data fusion and edge computing techniques to summarize and compress voluminous sensor data, thus reducing the energy cost of data transmission.
- Integration with Cloud-based analytics platforms to support post-mission analysis and long-term data management.

Conclusion

We have discussed system architecture and the various ways of breaking it up in terms of layers. We have examined and compared the well-known middleware, frameworks and standards used in AUV control architecture. Protocol stacks for how various layers of the system architecture interact have been discussed. Control strategies at the high-level and the low-level have been elaborated with examples of guidance and decision-making. The concepts of GNC for AUVs have been revisited and linked with the middleware and control strategies discussed.

Future research will focus on further enhancing autonomy, improving the transparency of automated decision-making processes, and developing scalable interfaces for managing large swarms of AUVs.

We have seen that effective user interfaces for AUV mission control must support situational awareness, minimise operator workload, prioritise critical data, and offer transparency and intervention capabilities, all of which contribute to increased trust and better overall mission performance.

In conclusion, the successful integration of human-machine interfaces, energy management strategies, and robust data processing systems is vital for the next generation of AUV missions. As the technology continues to evolve, the combined efforts of researchers and industry practitioners will lead to more efficient, reliable, and user-friendly systems that can operate effectively in the challenging underwater environment. The continuous feedback loop between operational data, human intervention, and automated system responses will ultimately enhance both mission safety and overall performance.

Having completed our journey through various components that go into designing, making and breathing the life of autonomous operation into UUVs, we now turn our attention to their operations in the real world. We have seen the range of applications for UUVs in the opening chapter; in the next chapter we take a closer look at their concepts of operations in specific missions.

References

Abdullah, A., Chen, R., Blow, D., & Uthai, T. (2024, December). Human-Machine Interfaces for Subsea Telerobotics: From Soda-straw to Natural Language Interactions. *eprint arXiv*:2412.01753. doi:DOI:10.48550/arXiv.2412.01753

Antonelli, G. (2018). *Underwater Robots.* Springer. doi:10.1007/978-3-319-72124-8

Benjamin, M.R., Schmidt, H., Newman, P.M., & Leonard, J. J. (2010). Nested autonomy for unmanned marine vehicles with MOOS-IvP. *Journal of Field Robotics, 27*, 834–875. doi:doi: 10.1002/rob.20370

Caccia, M. (2006). Autonomous surface craft: prototypes and basic research issues. *Autonomous Robots, 25*(4), 341-357.

Costanzi, R., Fenucci, D., & Manzari, V. (2020, July). Interoperability Among Unmanned Maritime Vehicles: Review and First In-field Experimentation. *Frontiers in Robotics and AI, 7*. doi: Volume 7 - 2020 | https://doi.org/10.3389/frobt.2020.00091

De Barros, E., Pascoal, A., & De Sa, E. (2004, July). AUV dynamics: Modelling and parameter estimation using analytical, semi-empirical, and CFD methods. *IFAC Proceedings Volumes, 37*, 369-376. doi:10.1016/s1474-6670(17)31760-3

DeMarco, K., West, M.E., & Collins, T.R. (2011). An implementation of ROS on the Yellowfin autonomous underwater vehicle (AUV). *OCEANS 2011 MTS/ IEEE*, (pp. 1-7). Kona. doi: 10.23919/OCEANS.2011.6107001

Fossen, T. (1991). *Modelling and Control of Underwater Vehicles.* Trondheim, Norway: Marine Cybernetics.

González-García, J., Gómez-Espinosa, A., Cuan-Urquizo, E., García-Valdovinos, L. G., Salgado-Jiménez, T., & Cabello, J. A. (2020). Autonomous Underwater Vehicles: Localization, Navigation, and Communication for Collaborative Missions. *Applied Sciences,* 10(4). doi:https://doi.org/10.3390/app10041256

Huang, H.-M. (2008). *Autonomy Levels for Unmanned Systems (ALFUS) Framework.* National Institute of Standards and Technology. NIST Special Publication 1011-I-2.0. Retrieved August 2024 from https://www.govinfo.gov/content/pkg/GOVPUB-C13-cbc9faa25f6d651e046c9df607d40d59/pdf/GOVPUB-C13-cbc9faa25f6d651e046c9df607d40d59.pdf

JAUS Mobility Service Set AS6009A. (2023, 10 6). Retrieved from SAE International. https://www.sae.org/standards/content/as6009a/

Moore, S., Bohm, H., & Jensen, V. (2010). *Underwater Robotics: Science, Design and Fabrication.* Monterey, CA: Marine Advanced Technology Education Center.

Oliveira, P., Pascoal, A., Silva, V., & Silvestre, C. (1998). Mission Control of the MARIUS AUV: System Design, Implementation, and Sea Trials. *International Journal of Systems Science,* 29(10), 1065-1080. doi:https://doi.org/10.1080/00207729808929598

Roberts, G.N., & Sutton, R. (eds.). (2013). *Further Advances in Unmanned Marine Vehicles.* The Institution of Engineering and Technology. doi:10.1049/PBCE002E

Soorati, M., Clark, J., Ghofrani, J., Tarapore, D., & Ramchurn, S. (2021). Designing a User-Centred Interaction Interface for Human-Swarm Teaming. *Drones,* 5(4), 131. doi:https://doi.org/10.3390/drones5040131

Whitsitt, S., & Sprinkle, J. (2011). Message modelling for the joint architecture for unmanned systems (JAUS). *18th IEEE International Conference and Workshops on Engineering of Computer-Based Systems,* (pp. 251–259). Las Vegas, NV. doi: 10.1109/ECBS.2011.17

Yoerger, D.R., Jakuba, M.V., Bradley, A.M., & Bingham, B. (2007). Techniques for Deep Sea Near Bottom Survey Using an Autonomous Robot. *International Journal of Robotics Research (IJRR),* 26(1), 41-54.

10

UUV Operations

The evolution of UUVs is transforming how navies execute missions for mine countermeasures, anti-submarine warfare (ASW), and intelligence, surveillance and reconnaissance (ISR). Today, unmanned systems are not just appendages controlled from the surface, but have become integrated, networked platforms that can operate with significant autonomy. Modern concepts of operations (CONOPS) for naval unmanned systems now emphasise not only increased operational range and endurance, but also capabilities such as loitering munitions and manned-unmanned teaming.

The first section of this chapter discusses concepts of operations of UUVs for naval roles, including key operational elements, control philosophies, and integration strategies. Examples are provided of AUVs deployed and demonstrated for naval operations.

The next section covers underwater surveys, which span industrial, scientific and defence applications. Parameters of interest, sensor technologies, and typical examples are outlined.

AUVs are essentially sensor nodes that capture data. Data quality and data management are of crucial importance for mission effectiveness and operations of AUVs. These aspects of data management, including data storage and transmission, are covered in the third section.

AUVs are most effective when their capabilities are multiplied through collaborative operations in swarms. However, these are limited by the challenges faced in acoustic communication. Several experiments and demonstration projects for swarm operations of AUVs and underwater sensor networks are described in the final section of the chapter.

CONOPS for Naval Unmanned Systems

The CONOPS (Concept of Operations) for any naval system describes how it is to be deployed, operated, and integrated into naval operations.[1] A well-developed CONOPS serves as a bridge between high-level strategic objectives and the detailed technical design of the system (in our case, UUVs).

Definition and Scope of CONOPS

In the naval context, a modern Concept of Operations (CONOPS) for naval unmanned systems would comprise several key elements, outlined below.

- **Mission Objectives:** What the unmanned systems are intended to accomplish (e.g., mine countermeasures, ISR, anti-submarine warfare, precision strike).

- **Operational Environment:** The types of conditions in which the systems will operate, including sea state, threat levels, communication limitations, and logistical support.

- **Roles and Responsibilities:** The allocation of tasks among unmanned platforms and between unmanned systems and manned vessels.

- **Command and Control Structure:** How decision making is organized, including the separation between high-level mission planning (backseat driver) and low-level control (front seat driver).

- **Interoperability Requirements:** How unmanned systems will integrate with other naval assets (e.g., aircraft, ships, submarines) and how data will be shared across the network.

- **Safety and Security Protocols:** Measures to ensure that the systems operate safely, reliably, and securely in contested or denied environments.

A robust CONOPS must be adaptable and evolve with emerging technologies and new threats. These can be divided into three phases: (a) Pre-mission planning and preparation, (b) Deployment and execution, and (c) Post-mission analysis and feedback.

---

[1]   Corfield & Hillenbrand, 2003

Pre-Mission Planning and Preparation
Objective Setting and Scenario Development

- Clearly define the mission objectives (e.g., mine countermeasures, ISR, ASW, loitering strike).
- Develop detailed scenarios including threat assessments, environmental conditions, and required operational timelines.

For example, an exercise might involve a mine countermeasure mission in a littoral zone where UUVs are tasked with scanning for mines and coordinating with a manned vessel.

Asset Allocation and Manned-Unmanned Teaming

- Determine the mix of unmanned and manned assets to be deployed.
- Define roles for each asset—for instance a manned submarine may serve as the command node while several AUVs perform reconnaissance and mine detection.
- Develop redundancy plans in case an unmanned system fails.

Communication and Data Management Protocols

- Plan for limited underwater bandwidth by using pre-defined data compression techniques and scheduled communication windows.
- Establish secure communication channels between unmanned systems and command centres.
- Define standard data protocols to ensure interoperability between systems from different manufacturers.

Deployment and Execution
Launch and Recovery Procedures

- Define the methods for launching unmanned systems (e.g., from a manned vessel, submarine, or unmanned surface vessel).
- Use specialised launch and recovery systems that minimize the risk to personnel and vehicles.

For example, an AUV like the REMUS 6000 or HUGIN 1000 is typically launched using an integrated recovery system from a surface ship, ensuring rapid deployment and safe recovery.

Autonomous Operation and Mission Management

- Implement advanced autonomous navigation and control algorithms. The 'backseat driver' approach allows high-level mission planning to run concurrently with low-level control loops.

- Allow for dynamic re-tasking based on real-time sensor inputs and environmental changes.

For example, a UUV may deviate from its planned path to avoid unexpected obstacles, and then rejoin the mission after passing the hazard.

Manned-Unmanned Teaming in Execution

- Use a distributed command structure where unmanned vehicles report real-time data to a manned vessel that serves as the operational hub.

- Enable cooperative behaviours among unmanned systems (e.g., swarm tactics) to cover large areas or track multiple targets.

For example, in a 2021 ASW trial, a manned destroyer coordinated with several UUVs to track and shadow an enemy submarine with each UUV autonomously adjusting its path based on shared sensor data.[2]

Loitering and Strike Capabilities

- For missions involving loitering munitions, define criteria for engagement (e.g., target identification, environmental thresholds).

- Ensure that UUVs designed for offensive roles have the necessary sensor suites and actuators to transition from reconnaissance to strike mode.

For example, a loitering UUV could remain in a designated area for up to 72 hours, monitoring enemy signatures before autonomously launching a precision strike.

Post-Mission Analysis and Feedback

Data Logging and Analysis

- Record detailed logs of vehicle performance, sensor data, and decision-making events.

- Use these logs to conduct after-action reviews to identify strengths, weaknesses, and areas for improvement.

- Integrate lessons learned into updated CONOPS and system designs.

---

[2]   Naval News Staff, 2021

Maintenance and Reconfiguration

- Schedule regular maintenance based on operational data and predicted component wear.

- Update software and hardware modules in a modular fashion, taking advantage of standardized interfaces.

- This iterative process ensures that unmanned systems remain state-of-the-art and mission capable.

Operational Reporting

- Provide feedback to strategic command on mission success, operational challenges, and potential threats identified during unmanned operations.

- Ensure that operational reports are standardized to facilitate comparison over time and between different missions.

Case Studies and Examples

CURV Series (1960s–1980s)

One of the earliest examples of unmanned underwater systems is the CURV (Cable-Controlled Underwater Recovery Vehicle) series (Figure 10-1).

Figure 10-1. The CURV ROV recovered a Hydrogen bomb lost in the sea off Palomares, Spain, in 1966[3]

---

[3]   (Sweeney, 1970), (CURV III, 2016)

Developed in the 1960s and refined through the 1970s and 1980s, CURV was used by the US Navy for salvage operations and early mine countermeasure tasks. Although it was a tethered system, its deployment during events such as the recovery of black boxes from downed aircraft and even the recovery of an unexploded Hydrogen bomb highlighted the potential of unmanned underwater operations.

REMUS Series (1990s–Present)

The REMUS (Remote Environmental Monitoring Units) series, developed by Woods Hole Oceanographic Institute (WHOI), exemplifies the evolution of AUVs.

- **REMUS 100 (1990s):** Designed for shallow-water missions, it was one of the first truly autonomous AUVs deployed for mine countermeasures and hydrographic surveys.
- **REMUS 600 (Late 1990s/Early 2000s):** Extended the operational depth and range, enabling deeper mine countermeasure and ASW missions.
- **REMUS 6000 (2000s–Present):** With a maximum depth capability of up to 6000 metres, the REMUS 6000 has been extensively used by the US Navy and allied navies in operational exercises.

This demonstrates the evolution of CONOPS of a scientific UUV from simple waypoint navigation to complex autonomous missions involving real-time decision making and manned-unmanned teaming for defence applications.

The 'Kingfish', specifically the MK 18 Mod 2, is a US Navy unmanned underwater vehicle (UUV) derived from the REMUS 600 platform. It is a medium-to-heavyweight UUV used for shallow water mine countermeasures and hydrographic reconnaissance. While the Kingfish is based on the REMUS 600, it features increased endurance and area coverage rate.[4]

---

[4]   Abott, 2023; Lundquist, 2021

**Figure 10-2. Kingfish UUV (Mark18 Mod 2) during
mine countermeasure operations in 2012**
(Source: US Navy[5])

Recent and Future Deployments

Over the last decade, concepts for integrated unmanned naval operations are becoming increasingly sophisticated, through steady increments in technology.

Manned–Unmanned Teaming

Modern CONOPS now frequently incorporate manned-unmanned teaming. The US Navy has conducted trials since 2015 for deploying AUVs from a manned submarine to conduct ASW missions. These trials envisaged the manned vessel to function as a command centre, processing data from multiple AUVs and directing their activities autonomously.

Loitering Munitions

In the past few years, research programs in Europe and the USA have begun to explore underwater loitering munitions. Although still in early stages, these systems are designed to remain on station for long periods, monitoring for targets, and then engaging when the opportunity arises. Such systems offer a dual role of ISR and precision strike, significantly altering naval CONOPS by enabling offensive operations with minimal human risk.

---

[5] Abott, 2023

Large Displacement UUVs (LDUUVs)

The US Navy and allied navies are developing LDUUVs capable of long-endurance missions. These vehicles are intended to operate in conjunction with manned vessels, forming a distributed network of sensors and strike assets. Their even-larger version, the Extra-Large UUVs (XLUUV), are expected to have mission endurance of 30 to 45 days, and capability to carry payloads of about 10 tons or more. Various countries have initiated programmes for their design and development in the past five years.

Swarm Operations

Emerging experiments have demonstrated that swarms of small UUVs can be coordinated to perform complex tasks such as large-area reconnaissance or mine detection. These systems rely on robust inter-vehicle communication and adaptive algorithms to function as a cohesive unit. These are covered further in a separate section in this chapter.

Deployment from Submarines

The first deployment of an AUV from a submarine Dry Deck Shelter was reported in July 2015 from the *Virginia* class SSN, *USS North Dakota* in the Mediterranean Sea.[6] The first AUV launch from a moving submarine was reported in August 2023.[7] Another *Virginia* class SSN (*USS Delaware*) made history in June 2025 by successfully launching as well as recovering an AUV through a torpedo tube without diver assistance. The *Yellow Moray* AUV used is a specialized version of the REMUS 600. During a series of three sorties lasting up to 10 hours each, the *Yellow Moray* completed a pre-programmed tactical objective, setting a technological milestone in undersea warfare.[8]

Underwater Surveys

The data parameters measured by AUVs through underwater surveys serve multiple industrial and scientific purposes. These include mapping and charting; geological and seismological studies; environmental monitoring; underwater archaeology and resource exploration. Underwater surveys have

---

[6]   Gady, 2015

[7]   Macey, 2023

[8]   Robertson, 2025

been discussed in chapter 1 when we looked at applications of UUVs. Here, we mention a few case studies and examples that illustrate successful deployments.

Hydrographic and Seafloor Mapping

A significant example of modern AUV capabilities is the record-breaking mission by Kongsberg's 'HUGIN Endurance' in 2024. In a multi-week autonomous mission, HUGIN Endurance covered a distance of more than 1,200 nautical miles. The AUV operated at depths ranging from 50 to 3,400 m, covering an area of approximately 36 km² with a resolution as high as 0.5 m. The mission demonstrated the vehicle's ability to operate independently with a positional error of only 0.02 per cent of the total distance travelled.[9]

The MARIUS AUV, developed in the mid-1990s, has been used extensively for coastal surveys and environmental assessments. With typical mission duration of around 18 hours, MARIUS has mapped coastal areas with a resolution of approximately 1.5 m and a maximum operational depth of 600/m. These surveys have provided data for environmental monitoring and resource management.

Environmental and Ecological Surveys

AUVs equipped with high-definition cameras and sidescan sonars have been used to map sensitive marine ecosystems such as coral reefs and seagrass beds. These surveys help in assessing habitat health, monitoring changes over time, and supporting conservation efforts. Data obtained from these missions have been used to generate 3D reconstructions of underwater habitats, providing valuable insights into ecological dynamics.[10]

Environmental monitoring missions have used AUVs to study sediment deposition, pollutant dispersion, and water quality parameters. For example, surveys using CTD sensors and fluorometers have helped researchers identify areas of high sedimentation or pollutant accumulation, guiding remediation efforts and policy decisions.

---

[9]   Kongsberg Discovery, 2024
[10]   Meduna, Petillot, & Lane, 2008

Industrial Application Surveys

AUV surveys are integral to offshore oil and gas exploration. Platforms like the EvoLogics BOSS and HUGIN 1000 have been used to map potential drilling sites, inspect underwater pipelines, and assess seafloor conditions. Data collected has resolutions ranging from 0.5 to 1.0 m, providing inputs for structural assessments and hazard evaluations.

AUVs have also been deployed for the inspection of marine infrastructure such as wind farm foundations, subsea cables, and port facilities. These missions use a combination of high-resolution sonar and visual cameras to detect structural defects, corrosion, or biofouling. The timely and accurate data provided by AUVs help reduce downtime and maintenance costs.

Table 10-1 provides a comparison of typical survey missions conducted by various AUV platforms. The data include the survey year, area covered, resolution, maximum depth, and mission duration.

Table 10-1. Surveys reported by typical AUV platforms

AUV Model	Year	Area Covered (km²)	Resolution (m)	Max Depth (m)	Mission Duration (h)	Application
HUGIN 1000	2000	5	1.0	3000	12	Hydrographic survey, oil exploration
MARIUS AUV	1996	8	1.5	600	18	Coastal mapping, environmental survey
SeaExplorer 3000	2018	10	1.0	2500	24	Seafloor mapping, scientific research
EvoLogics BOSS	2020	20	0.8	3500	30	Deep-water exploration
HUGIN Endurance	2024	36	0.5	3400	360 (15 days)	Long-range autonomous survey

The trend seen is towards increasing survey area and improved resolution in newer platforms. Mission durations have also expanded, from a few hours to multi-day operations, reflecting advances in battery technology and autonomy. Improvements in sensor technology, on-board processing, and energy management have enabled AUVs to cover larger areas with greater accuracy and at greater depths than ever before. These capabilities have made AUVs valuable tools for both industrial applications—such as pipeline inspection and resource exploration—and scientific research, including ecological monitoring and geological studies.[11]

Successful applications have also shown the importance of integrating advanced data management solutions. For instance, the HUGIN Endurance mission not only demonstrated impressive survey capabilities but also highlighted the need for robust on-board data fusion and adaptive transmission protocols to ensure that high-resolution data is reliably delivered despite energy constraints and communication challenges.

Data Management

The vast volumes of data generated during underwater surveys require robust data management systems. This section discusses how data are acquired, processed, transmitted, stored, and analysed. Data management solutions must address the challenges of on-board storage limitations, real-time processing constraints, and the need for robust post-mission data analytics.

Data Quality and Usefulness

The usefulness of underwater survey data depends on factors such as:

- **Resolution and Accuracy:** High-resolution data allow for detailed mapping and precise analysis, but may require more energy and bandwidth to acquire and transmit.

- **Coverage Area:** Extensive coverage is essential for large-scale assessments, though there is often a trade-off between area coverage and resolution.

- **Data Fusion:** Combining data from multiple sensors (e.g., sonar, video, and environmental sensors) enhances the overall quality and interpretability of the survey data.

- **Timeliness:** Real-time or near-real-time data are critical in dynamic environments or during emergency response scenarios.

Overall, high-quality survey data enable better decision-making in both scientific research and industrial applications, improve our understanding of the underwater environment and guide effective resource management and infrastructure development.

---

[11] Moore, Bohm, & Jensen, 2010

Data Acquisition and On-Board Storage

AUVs are equipped with an array of sensors, and the data generated can be extremely large. Typical data management solutions at the vehicle level are described below.

Sensor Data Characteristics

AUV sensors typically generate data with the following characteristics:

- **High Resolution and Volume:** High-frequency sensors (e.g., SAS, MBES) can produce gigabytes of data per mission.

- **Heterogeneous Data:** Data types include sonar imagery, video feeds, environmental measurements, and navigational telemetry.

- **Real-Time Requirements:** Certain data (e.g., navigation and system health) must be processed and transmitted in real time, while other data may be stored for later analysis.

On-Board Storage Solutions

AUVs incorporate on-board storage solutions to temporarily hold raw data until it can be offloaded:

- **Solid-State Drives (SSD):** Ruggedized SSDs are commonly used due to their high reliability and resistance to vibration and pressure.

- **Redundancy and RAID Configurations:** Multiple storage units are often configured in RAID to ensure data integrity even in case of hardware failure.

- **Compression Algorithms:** Data are often compressed on-board using both 'lossless' and 'lossy' techniques to maximize storage efficiency. Compression reduces the amount of data that must be transmitted, conserving both energy and bandwidth.

The integration of these storage systems is critical for ensuring that no data are lost during long missions, especially when real-time communication is not possible.

On-Board Processing and Edge Computing

Due to the constraints of underwater communication, AUVs need to perform significant on-board processing.

On-board processing capabilities enable AUVs to analyse and summarize data in real time. Real-time data processing (or 'Edge Computing') mentioned above is a pre-processing step that reduces energy consumption and allows for near real-time reporting of mission-critical data. Several techniques for 'data summarisation' help condense voluminous sensor data.

Deploying edge computing solutions allows complex computations to be performed on-board, reducing the volume of data that needs to be transmitted. This is also referred to as 'local decision making'. For example, machine learning algorithms can detect anomalies in sonar readings or classify objects in video feeds.

Some of the ways for data summarisation through real-time data processing are mentioned below. These summarization techniques are vital for both real-time reporting and long-term storage. They ensure that operators receive the most relevant information without being overwhelmed by raw data.

- **Data Fusion:** Data from various sensors are fused to produce a coherent representation of the underwater environment. For example, sonar and video data may be combined to enhance object detection. The AUV can generate higher-level abstractions (e.g., environmental maps) that require less bandwidth to transmit.

- **Adaptive Sampling:** Algorithms (on-board) can adjust the sampling rate of sensors based on environmental conditions. When the environment is stable, the sampling rate may be reduced to conserve energy.

- **Feature Extraction:** Algorithms can extract key features (e.g., edges, contours, statistical summaries) from high-resolution images and sonar data.

- **Anomaly Detection:** By focusing on anomalous data—such as sudden changes in sonar echoes or temperature spikes—the AUV can prioritize critical data for immediate reporting.

- **Dimensionality Reduction:** Techniques such as principal component analysis (PCA) or auto-encoders can reduce the dimensionality of the data, making it easier to store and transmit.

Data Transmission Strategies

We discussed underwater communication in chapter 7. Due to the limited bandwidth of underwater communication channels, data transmission strategies are critical. These comprise communication protocols and 'energy-aware' transmissions.

Communication Protocols

Underwater acoustic modems are the primary means of communication, though they offer low data rates. Protocols must be robust against noise and multipath effects.

Store-and-Forward: In situations where real-time data transmission is not feasible, AUVs may store data and offload it once a high-bandwidth link (e.g., when surfaced) is available.

Hybrid Communication: When AUVs surface or communicate via tethered buoys, higher-bandwidth radio frequency (RF) or satellite communications can be used to transmit detailed datasets.

Energy-Aware Transmission

- **Dynamic Modulation:** Adaptive modulation schemes that adjust transmission parameters based on current energy levels and channel conditions help optimize power usage.

- **Event-Triggered Reporting:** Instead of continuous data streaming, systems often use event-triggered protocols where only critical data (or summaries) are transmitted in real time.

- **Scheduled Data Bursts:** Data may be stored on-board and transmitted in bursts during predefined time windows when energy and communication conditions are optimal.

- **Prioritization of Critical Data:** Data are prioritized based on importance—critical alerts and diagnostics are transmitted immediately, while less urgent data are buffered.

Post-Mission Data Management

After each mission, large datasets must be offloaded and processed. Several options are available depending upon the volume, nature and time-criticality of the data.

Centralized or Distributed Storage

Data can be uploaded to central servers or Cloud platforms where robust backup, search, and retrieval systems ensure data integrity.

In other cases, data management may involve distributed storage across multiple geographic locations, which is particularly useful for multinational projects or long-term environmental monitoring.

Big Data Analytics

The enormous volume of data generated by AUVs calls for advanced analytical techniques.

- **Big Data Frameworks/Cloud-Based Processing:** Once data are offloaded, Cloud platforms provide the computational power required to analyze and interpret large datasets. Frameworks such as Hadoop and Spark are used for data mining and trend analysis.

- **Machine Learning:** Advanced algorithms are applied to the data to detect patterns, classify features, and predict future environmental changes. This analysis is critical for refining future survey missions and improving the design of AUVs.

- **Visualization Tools:** Data visualization software transforms complex datasets into intuitive graphics—such as 3D maps, heat maps, and temporal trend charts—allowing scientists and engineers to extract actionable insights.

Data Fusion and Integration

Data fusion techniques integrate data from multiple sensors to provide a comprehensive picture of the underwater environment. Effective data fusion can reduce redundancy, improve accuracy and facilitate better decision-making.

By combining overlapping datasets, the system can eliminate redundancy and focus on the most relevant information. Fusing data from different sensors (e.g., sonar and video) often leads to more accurate and robust measurements. A coherent, fused dataset helps operators make informed decisions during mission execution and in post-mission analyses.

Integrating data over time allows for trend analysis and improves the accuracy of environmental models. This is particularly useful in dynamic environments where conditions change rapidly.

In summary, robust data management solutions in AUV operations encompass on-board processing and summarization, adaptive transmission protocols, centralized and distributed data storage architectures, and advanced big data analytics. Together, these solutions enable efficient handling of large data volumes while supporting real-time operational requirements and post-mission analysis.

Collaborative AUV Operations

Motivation

In many underwater missions—such as seabed mapping, environmental monitoring, or surveillance—a single AUV may take many hours or days to complete its task. In contrast, a coordinated swarm can divide the area of interest into sectors, each covered by an individual agent. This division of labour not only reduces mission duration but also increases spatial resolution and data redundancy. The major motivations for collaborative AUV operations are discussed further.

Enhanced Coverage and Efficiency: By using multiple agents that can share data in real time, it becomes possible to detect and localise anomalies (e.g., pollutant plumes or underwater hazards) more accurately than would be feasible with a solitary unit. In addition, coordinated path planning algorithms allow the swarm to adapt dynamically to new information, redistributing agents as needed to areas of interest.

Robustness and Redundancy: Swarm operations inherently offer greater resilience compared to single-vehicle systems. In a well-designed swarm, the failure or temporary malfunction of one or a few agents does not compromise the overall mission. Instead, the remaining vehicles can reconfigure their trajectories and roles to compensate for the loss. This redundancy is especially critical in harsh underwater environments where reliability is paramount and recovery operations may be expensive or impractical.

Adaptability and Real-Time Decision-Making: The distributed nature of swarm control allows individual agents to make decisions based on local sensor data, while still contributing to a global mission objective. Advances in machine learning and decentralized control algorithms have enabled AUVs to perform tasks such as obstacle avoidance, target tracking, and dynamic re-tasking without constant human supervision. This level of autonomy is key to adapting to rapidly changing underwater conditions (such as variable currents, obstacles, or communication disruptions) and ensures that the swarm can operate effectively in complex and unstructured environments.[12]

Scalability and Cost-Effectiveness: Swarm operations are typically based on modular designs where relatively low-cost vehicles work in concert. This modularity makes it economically feasible to deploy large numbers of AUVs. In contrast to high-end systems that are expensive to maintain, a swarm composed of inexpensive, mass-produced vehicles can achieve significant coverage and redundancy. Moreover, advances in miniaturized sensors and energy-efficient designs further contribute to the scalability of these systems.

Swarm Operations

Collaborative AUV operations have been the focus of extensive research and several experimental projects worldwide. We now look at some notable efforts and experimental results.

SWARMs

The SWARMs project (Swarm Autonomous Robotic Marine Systems) has been one of the prominent European initiatives in this area.[13] Funded under a European framework program, the project focused on following aspects:

- Developing decentralized control architectures.
- Implementing real-time data fusion and collaborative mapping.
- Demonstrating field trials with up to eight cooperating AUVs.
- Addressing challenges in underwater communication using hybrid acoustic and optical methods.

AQUASwarm

AQUASwarm is another European initiative that emphasized environmental monitoring and mapping using AUV swarms. Key aspects of AQUASwarm include:

- Adaptive mission planning based on real-time data.
- Energy-efficient operations designed for extended endurance.
- Field experiments with 5–6 AUVs operating in coastal environments.

---

[12] Hambling, 2016

[13] Real-Arce, 2016

- Demonstrated improvements in mapping resolution and mission speed.

MARUS Collaborative Experiments

European research groups associated with the MARUS consortium in Germany have conducted collaborative experiments that demonstrate:

- Cooperative navigation and formation control in coastal waters.
- Multi-hop communication techniques that extend the effective operational range of the swarm.
- Integration of sensor data from multiple AUVs for improved situational awareness.
- Robustness of the swarm in environments with complex underwater terrain and variable currents.

Other European projects and consortia have contributed through workshops, technology demonstrations, and collaborative publications that advance the theoretical and practical aspects of AUV swarm operations. These efforts provide a strong foundation for future deployments in both civilian and military domains. Table 10-2 summarises key experimental projects of collaborative AUV operations, highlighting parameters demonstrated.

Table 10-2. Examples of Collaborative AUV Projects (Swarm operations) demonstrated

Project Name	Agents	Operations Area/ Distance	Endurance	Communi-cation Method	Notable Characteristics
SWARMs (2016–2019)	8	Up to 10 km² area; 1–2 km range (acoustic)	~24 hours	Hybrid (Acoustic, Optical)	Real-time data fusion, decentralized control, adaptive mission planning
AQUA Swarm (2017–2019)	5–6	3–5 km coastal operations	~20 hours	Primarily Acoustic	Energy-efficient operations, real-time adaptive mission management
MARUS (2015–2018)	4–5	4 km operational range	~18 hours	Acoustic modems (multi-hop)	Formation control, cooperative obstacle avoidance, robust navigation
EuRo-AUV (2018–2020)	6	~6 km operational distance	~22 hours	Acoustic + Satellite relay	Multi-mission adaptability, integrated sensor networks
Cooperative Ocean Survey (COOS) (2019–2021)	7	5–7 km area coverage	~20–24 hours	Acoustic and RF (near-surface)	High-resolution mapping, adaptive reconfiguration, redundant sensing

Underwater Networks

Internet of Underwater Things (IoUT)

The future of collaborative undersea operations would involve networked sensor and vehicle systems, referred to as Internet of Underwater Things (IoUT) or Subsea Internet of Things (SIoT).[14] Such underwater networks would need to rely largely on acoustic communication. In the earlier chapters, we mentioned the relevance of IoUT for optimising AUV energy consumption and the significance of underwater communication for realising this concept. Initial efforts in this area were by the US Office of Naval Research (ONR) and DARPA through the PLUSNet (Persistent Littoral Undersea Network) project that started in 2005.[15]

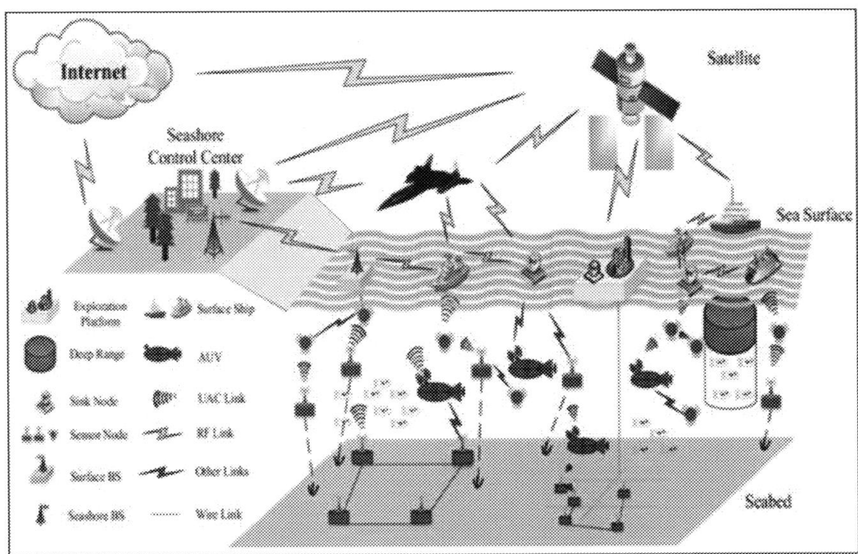

Figure 10-3. Model of IoUT with various types of communication networks[16]

SeaWeb

In an article entitled 'Enabling Undersea ForceNET with Seaweb Acoustic Networks' in 2003, Joseph Rice of SPAWAR San Diego predicted: "Undersea, off-board, autonomous systems will enhance the war-fighting effectiveness of submarines, maritime patrol aircraft, amphibious forces, battle groups, and space satellites. Wide-area sensor grids, leave-behind multi-static sonar sources,

---

[14] Nkenyereye, Nkenyereye, & Ndibanje, 2024

[15] Stewart & Pavlos, 2006

[16] Qiu, Zhao, Zhang, Chen, & Chen, 2020

mine-hunting robots, and AUVs are just a few of the battery-powered, deployable devices that will augment space and naval platforms."[17]

In 2010, researchers at the US Naval Postgraduate School in Monterey, California, unveiled 'Seaweb' after more than ten years of development. Seaweb is an acoustic communications network which uses through-water acoustic modems to connect up with a large number of sensor nodes to a surface gateway node, allowing real-time digital communication between data gathering sensors and land-based command centres.[18]

Their original goal was to create a network of distributed sensors for detecting quiet submerged submarines in littoral waters where traditional ASW surveillance is challenged by complex sound propagation and high noise. In this concept, communication between non-tethered UUVs and submarines are provided via underwater digital acoustic communications (ACOMMS). ACOMMS is a short-range, modem-like communication method that is often left unencrypted (though it could easily be encrypted). [19]

In 2005, SeaWeb ran three experiments using their expendable node technology to aid in the navigation of several types of UUVs. These tests showed that undersea ACOMMS technologies using multiple scattered nodes are a viable solution for naval communication. The next step for Seaweb was to utilize the sensor nodes to communicate to and from UUVs and submarines, as well as relay mission specific information to and from satellites via various surface assets, as depicted in Figure 10-4.

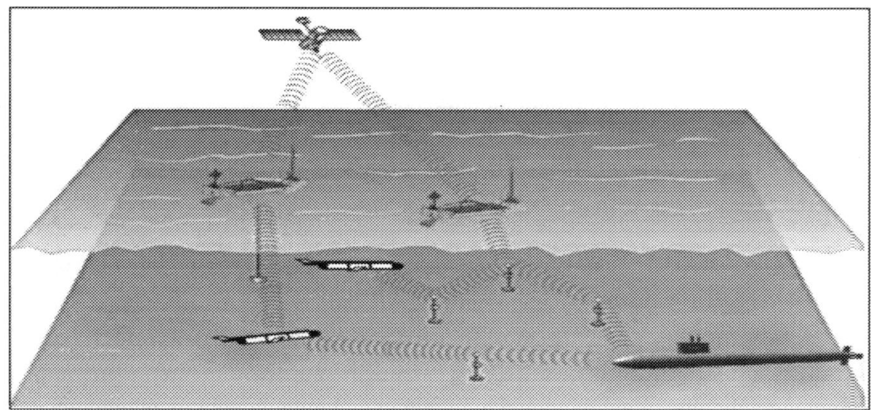

Figure 10-4. Representation of 'SeaWeb' acoustic communication and navigation network[20]

[17]   *Ibid.*
[18]   Lo, 2012
[19]   Vandenberg, 2010
[20]   Rice & Green, 2008

PLUSNet

PLUSNet was an applied research program funded by the Office of Naval Research from 2005, integrating more than a dozen demonstrated technologies to test and evaluate the merits of a prototypical, persistent, automated, tactically and environmentally adaptive sensor network.

Figure 10-5. Concept of Persistent Littoral Undersea Network (PLUSNet) of ONR & DARPA[21]

This network of fixed and mobile sensors and underwater communications nodes was to employ advanced low-power signal processors for contact detection, classification, and localization. It would exercise different levels of autonomy from a remote central controller, including contact reporting, collaborative tracking, and contact hand-off to an adjacent sensor network. It was envisaged to demonstrate coupling of a submarine with this off-board

---

[21] Martin, 2005

sensor network (consisting of mobile AUVs as well as fixed sensors) to develop a persistent ASW capability against quiet diesel submarines in shallow waters.[22]

Prospects and Research Challenges

While significant progress has been made in collaborative AUV operations, several research challenges and future directions remain, which are touched upon here.

Decentralized Control Algorithms

A major area of research has focused on developing decentralized control algorithms that enable individual AUVs to operate collaboratively without relying on a central command. Techniques such as consensus algorithms, behaviour-based robotics, and artificial potential fields have been adapted for underwater applications. These algorithms allow agents to share information about their positions, velocities, and sensed data with neighbours, thus forming a dynamic network that can achieve coordinated behaviour. Laboratory experiments and field trials have demonstrated that such algorithms can ensure robust formation-keeping, adaptive reconfiguration, and obstacle avoidance even in the presence of environmental disturbances.

Communication Protocols

Underwater communication is notoriously challenging due to the limitations of acoustic channels, such as low data rates, high latency, and signal attenuation. Researchers have worked on developing communication protocols that are robust to these issues. Recent work has seen the integration of hybrid communication approaches—combining acoustic, optical, and even radio frequency (RF) links (when near the surface)—to facilitate inter-agent communication. Experimental tests have shown that with proper scheduling and error correction, swarms can maintain effective communication over several kilometres in complex underwater environments.[23]

Robustness under Harsh Conditions

Future research must address the robustness of swarm operations in highly dynamic and uncertain underwater environments. This includes:

---

[22]   Stewart & Pavlos, 2006
[23]   Roberts & Sutton, 2013

- **Resilience to Communication Loss:** Developing protocols that allow the swarm to continue operating effectively even if some agents temporarily lose connectivity.

- **Fault Tolerance:** Improving redundancy and self-healing algorithms so that the swarm can automatically compensate for the failure or degradation of one or more agents.

Enhanced Autonomy and Intelligence

Advances in artificial intelligence and machine learning are expected to further enhance the autonomy of AUV swarms. Areas of focus include:

- **Predictive Analytics:** Utilizing historical mission data to predict environmental changes and adjust mission parameters proactively.

- **Decentralized Learning:** Enabling agents to learn from their collective experiences in a decentralized manner, thereby improving overall mission performance over time.

- **Hybrid Swarm Architectures:** Integrating heterogeneous platforms (e.g., AUVs, gliders, and unmanned surface vehicles) to leverage the strengths of different vehicle types for comprehensive mission coverage.[24]

Energy Management and Endurance Optimization

Energy remains a critical constraint in underwater operations. Research is needed on the following aspects affecting operations.[25]

- **Optimize Energy Consumption:** Develop more efficient propulsion systems, sensor packages, and control algorithms that reduce overall power consumption.

- **Incorporate Energy Harvesting:** Integrate underwater energy harvesting technologies (e.g., from wave or thermal gradients) to extend mission endurance.

- **Smart Power Distribution:** Implement intelligent power management systems that allocate energy resources optimally across the swarm.

---

[24]   Colmer & Ellem, 2021
[25]   Griffiths, 2003

Interoperability and Standardization

The success of collaborative operations depends on the ability of diverse platforms to work together seamlessly. Future efforts should aim to:

- **Establish Standard Protocols:** Develop internationally recognized standards for underwater communication, data sharing, and control interfaces.

- **Promote Modular Design:** Encourage the use of modular hardware and software architectures that facilitate interoperability among systems developed by different organizations.

Integration with Surface and Space Assets

The integration of underwater swarms with surface vessels, satellites, and aerial platforms offers exciting new opportunities for mission management and real-time data relay. Such integrated systems could potentially overcome some of the limitations of underwater communication and provide a seamless link between the underwater environment and command centres on land.

Collaborative operations using swarms of Autonomous Underwater Vehicles represent a transformative approach to underwater missions. By leveraging the strengths of distributed sensing, decentralized control, and cooperative data fusion, AUV swarms offer the potential to achieve unprecedented coverage, resilience, and adaptability. We have discussed some visionary but ambitious goals of networking not only underwater nodes but linking them to assets on the surface, in the air, and in space.[26] If realised, these could truly be 'force multipliers' in the maritime domain.[27]

Way Ahead

Operational objectives are the purpose for which any platform is designed and developed. This chapter has covered various aspects and examples of UUV operations for defence applications, followed by examples of underwater surveys. We have discussed the importance and approaches for data management, which includes data acquisition, storage, transmission and integration.

We have seen collaborative UUV operations from two facets: swarm operations (with examples of scientific research projects) and sensor networks

---

[26] Nichols, Mumm, et al, 2020

[27] Ray, Singh, & Seshadri, 2011

(with examples of defence research projects). Prospects and research challenges for collaborative UUV operations have been mentioned, which can point the enthusiastic engineer to potential areas of research.

Having covered various operational aspects in this chapter, the next chapter moves to the final stages of 'getting your feet wet'. We will cover the mechanical solutions for launching UUVs from the ship into the water, and how their trials are planned and scheduled before acceptance or delivery.

References

Abott, R. (2023, July 2). *MK 18 Mod 2 Underwater Drone Program Production Finished.* Retrieved from Defense Daily: https://www.defensedaily.com/mk-18-mod-2-underwater-drone-program-production-finished/navy-usmc/

Bluefin-21 Autonomous Underwater Vehicle (AUV). (2014, April 14). Retrieved from *Naval Technology*: https://www.naval-technology.com/projects/bluefin-21-autonomous-underwater-vehicle-auv/

Colmer, A., & Ellem, R. (2021). *Remote Undersea Surveillance.* Defence Science and Technology Group, Emerging Disruptive Technology Assessment Symposium (EDTAS). Australian Government Department of Defence. Retrieved March 2024 from https://www.dst.defence.gov.au/sites/default/files/events/documents/Insights%20Paper%20-%20Remote%20Undersea%20Surveillance%20F1.pdf

Corfield, S., & Hillenbrand, C. (2003). Defence Applications for Unmanned Underwater Vehicles. In G. Griffiths (ed.), *Technology and Applications of Autonomous Underwater Vehicles* (pp. 161-178). London: CRC Press. doi:10.1201/9780203522301.ch10

CURV III. (2016, April 21). Retrieved from United States Naval Undersea Museum: https://navalunderseamuseum.org/curv-iii/

Davis, M. (2021, March 24). *Robots at war: The future for autonomous systems at sea in the Indo-Pacific.* Retrieved from ORF Online: https://www.orfonline.org/expert-speak/robots-war-future-autonomous-systems-sea-indo-pacific

Gady, F.-S. (2015, July 24). *Confirmed: US Navy Launches Underwater Drone From Sub.* Retrieved from *The Diplomat*: https://thediplomat.com/2015/07/confirmed-us-navy-launches-underwater-drone-from-sub/

Geyer, R. (1977). *Submersibles and their Use in Oceanography and Ocean Engineering.* New York: Elsevier.

Griffiths, G. (ed.). (2003). *Technology and Applications of Autonomous Underwater Vehicles.* Taylor & Francis. doi:10.1201/9780203023249

Hambling, D. (2016, March). The Inescapable Net: Unmanned Systems in Anti-Submarine Warfare. *BasicInt.* Retrieved from https://basicint.org/publications/david-hambling/2016/inescapable-net-unmanned-systems-anti-submarine-warfare

Kongsberg Discovery. (2024, September 4). *Smashing AUV records, HUGIN Endurance completes a multi-week fully autonomous mission.* Retrieved May 2025 from Kongsberg.com: https://www.kongsberg.com/newsroom/news-archive/2024/smashing-auv-records-hugin-endurance-completes-a-multi-week-fully-autonomous-mission/

Lo, C. (2012, April 30). *Persistent littoral surveillance: Automated coast guards.* Retrieved August 2024 from https://www.naval-technology.com/features/featurenavy-persistent-littoral-surveillance-auvs-uuvs/?cf-view

Lundquist, N. (2021, February 15). *Subsea Defense: Navy Deepens Commitment to Underwater Vehicles.* Retrieved from *Marine Technology News*: https://www.marinetechnologynews.com/news/subsea-defense-deepens-commitment-608367

Macey, J. (2023, July 28). *AUV Torpedo Tube Launch & Recovery Solution from Underway Submarine.* Retrieved from *Defense Advancement*: https://www.defenseadvancement.com/news/auv-torpedo-tube-launch-recovery-solution-from-underway-submarine/

Martin, D. L. (2005). Autonomous Platforms in Persistent Littoral Undersea Surveillance: Scientific and Systems Engineering Challenges. NPS Menneken Lecture Series.

Meduna, Z., Petillot, Y., & Lane, D. (2008). Long-term autonomous underwater vehicle navigation using terrain matching. *IEEE Journal of Oceanic Engineering, 33*, 103–121.

Moore, S., Bohm, H., & Jensen, V. (2010). *Underwater Robotics: Science, Design and Fabrication.* Monterey, CA: Marine Advanced Technology Education Center.

Naval News Staff. (2021, January 13). *ASW Exercise Sea Dragon 2021 Kicked Off with US Navy, RAAF, RCAF, Indian Navy and JMSDF.* Retrieved from Naval News: https://www.navalnews.com/naval-news/2021/01/asw-exercise-sea-dragon-2021-kicked-off-with-us-navy-raaf-rcaf-indian-navy-and-jmsdf/

Nichols, R. K., Mumm, H. C. et al. (2020). *Unmanned Vehicle Systems & Operations on Air, Sea, Land.* New Prairie Press. doi:10.4148/2334-0402.1030

Petillot, Y. R., Antonelli, G., Casalino, G., & Ferreira, F. (2019, June 2). Underwater Robots: From Remotely Operated Vehicles to Intervention-Autonomous Underwater Vehicles. *IEEE Robotics & Automation Magazine, 26*, 94-101. doi:10.1109/MRA.2019.2908063

Qiu, T., Zhao, Z., Zhang, T., Chen, C., & Chen, C. (2020). Underwater Internet of Things in Smart Ocean: System Architecture and Open Issues. *IEEE Transactions on Industrial Informatics, 16*, 4297–4307.

Ray, A., Singh, S., & Seshadri, V. (2011). Underwater Gliders: Force Multipliers for Naval Applications. *Warship 2011: Naval Submarines and UUVs.* Bath, UK: Royal Institution of Naval Architects. doi:10.3940/rina.ws.2011.03

Real-Arce, D. A., Morales, T., Barrera, C., Hernández, J., & Llinás, O. (2016). Smart and networking underwater robots in cooperation meshes: the swarms ECSEL: H2020 project. In *Instrumentation Viewpoint* (No. 19, pp. 19-19). SARTI.

Rice, J. A., & Green, D. (2008). Underwater Acoustic Communications and Networks for the US Navy's Seaweb Program. *Sensor Technologies and Applications SENSORCOMM '08.* San Diego. doi:10.1109/SENSORCOMM.2008.137

Roberts, G. N., & Sutton, R. (eds.). (2013). *Further Advances in Unmanned Marine Vehicles.* The Institution of Engineering and Technology. doi:10.1049/PBCE002E

Robertson, R. (2025, June 14). *US Navy makes history with unmanned underwater vehicle.* Retrieved from SAN (Straight Arrow News): https://san.com/cc/us-navy-makes-history-with-unmanned-underwater-vehicle/

Stewart, M. S., & Pavlos, J. (2006). A Means to Networked Persistent Undersea Surveillance.

Submarine Technology Symposium. Retrieved September 2024 from http://mseas.mit.edu/archive/PLUSNet/5stewamsPAF2.pdf

Sweeney, J. B. (1970). *A Pictorial History of Oceanographic Submersibles.* London: Robert Hale & Company.

Vandenberg, T. D. (2010). *Manning and Maintainability of a Submarine Unmanned Undersea Vehicle (UUV) Program: A Systems Engineering Case Study.* Monterey, CA: Naval Postrgraduate School. Retrieved January 2025, from https://apps.dtic.mil/sti/tr/pdf/ADA531594.pdf

11

Deployment and Trials

Any UUV gets tested in water, and ultimately, in the ocean. Getting a vehicle into the water, seeing it float and then allowing it dive out of sight is the ultimate test of confidence for its designers, developers and builders.

In this chapter, we examine the means for deploying UUVs into the sea, which are termed as 'Launch and Recovery Systems' (LARS). The beginning and end of every UUV mission are governed by this critical system. In case of manned submersibles, these are the evolutions that pose the maximum risk to personnel. In case of unmanned systems, these pose the maximum risk for loss of the expensive and complex vehicle. The first part of this chapter covers various types of LARS, their design considerations, and range of features.

Accounts of UUV development and operations are replete with challenges during testing.[1] A thoughtfully planned sequence of tests and trials is essential for time-bound development.[2] These need to be phased, and measurement parameters linked meaningfully to mission requirements. The second part of this chapter describes the preparation of a testing and evaluation plan, and a typical schedule for conducting UUV trials ashore, in harbour and at sea.

Launch and Recovery Systems

UUV Launch and Recovery Systems (LARS) are systems engineered to ensure the safe deployment and retrieval of UUVs from a support platform. These systems can be integrated into the hull or deck of a vessel, installed on dedicated platforms, or even be part of unmanned surface vehicle (USV) operations. Their primary function is to bridge the interface between the dynamic marine

---

[1]   Butler, 2018

[2]   Griffiths, Millard, & Rogers, 2003

environment and the relatively delicate underwater vehicle as it enters the water. LARS must ensure that vehicles are transferred through the splash zone safely, minimize the risk of damage or loss, and enable rapid turnaround.

Early shipboard systems for launching and recovering underwater vehicles used simple cranes or boat davits, which still are common solutions.[3] Even now, man-portable AUVs (less than say, 30 kg) can just be dropped over the side from a boat, but recovering them requires precise homing and is a bigger challenge.

Over time, as underwater vehicles became more complex and capable of deeper dives, LARS had to evolve. In the 1960s and 1970s, as ROVs and AUVs began to be used for tasks such as cable surveys and underwater salvage, specialized systems were developed that featured mechanisms such as A-frames and telescopic arms.[4] Deployment of ROVs have challenges related to handling of a long umbilical mounted on a winch, which needs to paid out (and recovered) for hundreds (or even thousands) of metres.[5]

Organizations such as the Woods Hole Oceanographic Institution (WHOI) and companies like NOV-BLM led the way by designing integrated systems that could operate in harsh sea conditions and support vehicles with increasing payload and performance requirements (Figure 11-1).[6]

Figure 11-1. A Launch and Recovery System for UUV
(Source: WHOI)[7]

---

[3]   Ballard, 2021

[4]   Rumson, 2021

[5]   Pantheeradiyil, 2024; Christ, 2011.

[6]   NOV, 2025; WHOI.

[7]   WHOI

Challenges in Launch and Recovery

LAR Systems face challenges due to the nature of the marine environment and the operational requirements of underwater vehicles. These challenges can be categorised into environmental, operational, technical, and human factors.

Environmental Challenges

Sea State and Weather Conditions

The most significant challenge for any LARS is the unpredictable and often severe marine environment. High sea states characterised by large waves, strong currents, and turbulent conditions make it difficult to maintain precise control over the deployment and recovery process. For instance, a system designed for operations in sea state 2 may not be reliable in sea state 5. Therefore, LARS must be robust enough to function in a range of environmental conditions while ensuring the safety of both the UUV and the support vessel (Figure 11-2).

**Figure 11-2. Israeli AUV Blue Whale being launched by
German Navy during trials in Nov 24**
(Source: German MoD)[8]

---

[8]　https://armyrecognition.com/news/navy-news/2025/greece-to-use-israeli-bluewhale-unman
ned-submarine-to-improve-maritime-intelligence-gathering

Dynamic Vessel Motion

Support vessels are seldom stationary, particularly during operations at sea. The motion induced by waves, tides, and wind means that the launch and recovery system must accommodate significant heave, pitch, and roll. A well-designed LARS will include heave compensation mechanisms such as shock absorbers, telescopic arms, or dynamic winch systems that can adapt to these movements in real time, ensuring that the vehicle remains stable during the transition through the water's surface (Figure 11-3).

Figure 11-3. AUV attached to LARS through motion compensation mechanism[9]

Hydrodynamic Forces

The process of lowering or hoisting a vehicle through the water subjects the system to complex hydrodynamic forces. Drag, lift, and turbulence can cause unwanted oscillations or misalignment, which in turn may damage sensitive instruments on the vehicle or the recovery system itself. Hydrodynamic design considerations are therefore critical; LARS designers use computational fluid dynamics (CFD) and physical scale model testing to optimize system shapes, materials, and motion-compensation features.

---

[9]   Sagar, Konara, Picard, & Park, 2025

Operational Challenges

Safety Considerations

Safety is paramount during launch and recovery operations. A malfunction or misalignment during these operations can result in damage to expensive vehicles, injury to personnel, or even catastrophic loss of life. Safety features in LARS often include fail-safe mechanisms, redundant systems, and real-time monitoring of load, tension, and position. Additionally, remote or autonomous systems are increasingly employed to reduce human exposure to hazardous conditions on deck during rough seas.

Load Management

Underwater vehicles vary widely in size and weight. The LARS must be capable of managing dynamic loads associated with heavy vehicles as well as lighter ones. This includes not only the static weight of the vehicle but also dynamic forces such as impact loads from wave action and inertial loads during acceleration and deceleration. The use of advanced winch systems, counterweights, and load-distribution mechanisms is common in modern LARS designs.

Integration with Vehicle and Vessel Systems

Launch and Recovery Systems do not operate in isolation; they must integrate seamlessly with both the underwater vehicle and the support vessel's systems. This involves compatibility with the vehicle's attachment points, communication protocols, and control systems. At the same time, the LARS must be integrated into the vessel's structure without compromising its stability or operational capabilities. Achieving this level of integration often requires engineering solutions customised to the specific vehicle and vessel combination. The ability to fit LARS to smaller surface ships (Figure 11-4) presents opportunities to operate UUVs from USVs.

**Figure 11-4. Sea Launcher LARS (by Hydroids)
deploying a REMUS 600 AUV[10]**

Technical Challenges

Mechanical Complexity

Modern LARS often incorporate sophisticated mechanical components such as telescopic arms, winches, cranes, and shock-absorbing cradles. These components must be precisely engineered to withstand the harsh conditions of the marine environment while performing reliably over many cycles. Mechanical failures such as cable fatigue, hydraulic leaks, or component wear, are significant concerns. As such, design approaches must include rigorous testing protocols and the use of high-grade, corrosion-resistant materials.

Control and Automation

Automation is increasingly common in LARS to improve safety and efficiency. However, developing control algorithms that can manage the complex interactions between vessel motion, environmental forces, and mechanical system dynamics is challenging. Feedback from sensors such as accelerometers, gyroscopes, GPS, and acoustic positioning systems must be integrated into a coherent control system. This system needs to adjust in real time to disturbances and unexpected events during launch and recovery.

---

[10]   https://bluezonegroup.com.au/announcements/remus-autonomous-launch-recovery/

Communication and Data Transfer

Effective communication between the LARS, the support vessel, and the underwater vehicle is critical, especially when systems are remotely operated. Underwater communications, often relying on acoustic modems, have limited bandwidth and can be affected by environmental noise. Ensuring reliable data transfer for control signals and video feedback requires robust communication systems and often the use of redundant channels.

Human Factors and Operational Procedures

The interface between human operators and automated systems is another critical challenge. Even with high levels of automation, human oversight remains essential during complex recovery operations. The design of the operator control interface, including clear visual and auditory feedback, is crucial to ensure that operators can intervene quickly if necessary. Training and simulation tools, such as virtual reality and physical mock-ups, are increasingly used to prepare crews for handling LARS in a range of conditions.

Categories of LARS Solutions

LARS solutions vary widely in design, reflecting differences in application, vehicle size, operational environment, and platform type. Broadly, these solutions can be categorized into four main types.

Fixed Platform Systems

Fixed platform systems are typically integrated into the deck or hull of a vessel. They include:

- **Crane-Based Systems:** These systems use shipboard cranes to lift the underwater vehicle directly from the water (Figure 11-5). They are common on larger vessels where deck space is not a limiting factor. The simplicity of a crane makes it a robust solution, although it may require additional mechanisms (e.g., shock absorbers or specialized slings) to manage dynamic loads.

Figure 11-5. Crane with padded hat and release/capture device to
minimise free swing angle, which maintains AUV attitude
(Source: MBARI[11])

- **A-Frame Systems:** An A-frame mounted on the stern or side of a vessel provides a guided structure for lowering and raising vehicles (Figure 11-6). The A-frame can help align the vehicle and protect it during the transition through the splash zone. Many systems incorporate heave compensation features to mitigate vessel motion.

Figure 11-6. LARS A-frame arrangement for deploying submersibles[12]

---

[11] Polar AUV Guide: Deployment and Recovery

[12] https://www.royalihc.com/offshore-energy/offshore-equipment/launch-and-recovery-systems

- **Telescopic Arms and Gantries:** These systems offer a compact solution that can extend over the water and retract once the vehicle is in the water. Telescopic designs are particularly useful in situations where deck space is limited (Figure 11-7).

Figure 11-7. Ramp-type LARS[13]

Mobile Vessel Systems

Mobile vessel systems are those integrated into support vessels that may not be dedicated solely to underwater vehicle operations. They include:

- **Dedicated Launch and Recovery Vessels (LRVs):** These are ships built specifically or heavily modified to support underwater operations. They often include integrated cranes, winches, and dynamic positioning systems to maintain precise location during operations.

- **Unmanned Surface Vehicles (USVs) with LARS:** Increasingly, autonomous or remotely operated surface vehicles are being used as mobile platforms for underwater vehicle launch and recovery (Figures 11-8, 11-9). USVs offer the advantage of reducing human risk and can operate in challenging conditions with advanced station-keeping capabilities.[14]

---

[13] https://okeanus.com/products/launch-and-recovery-systems/ramp-style-lars

[14] Thales, 2020

Figure 11-8. USV SAND (Surface Advanced Naval Drone) of
Fincantieri operating as mother platform for AUVs[15]

Figure 11-9. Henriksen LARS demonstrated for
HUGIN AUV from unmanned boat
(Source: H. Henriksen AS)[16]

- **Hybrid Vessel Systems:** These systems combine the features of manned
  and unmanned vessels to offer a flexible and cost-effective platform
  for underwater operations.

---

[15] Peruzzi, 2023

[16] Hydro International, 2020

Autonomous and Remote LARS

Autonomous systems aim to minimize or eliminate human intervention during launch and recovery:

- **Automated LARS on USVs:** In these systems, an unmanned surface vessel is equipped with an automated launch and recovery mechanism. For example, a system may use acoustic positioning and remote-controlled winches to synchronize the retrieval of an AUV.[17]

- **Robotic Docking Stations:** Some systems use fixed or mobile docking stations that allow underwater vehicles to 'dock' autonomously for battery recharging, data transfer, or maintenance (Figures 11-10, 11-11). These are particularly valuable in long-duration missions where frequent retrieval to the surface is not feasible.

The development of automated launch and recovery systems integrated into unmanned surface vehicles (USVs) involves the following aspects:

- **Autonomous Coordination:** The USV is equipped with a top-mounted winch and telescopic arm system that can lower an AUV into the water while the USV is in motion.

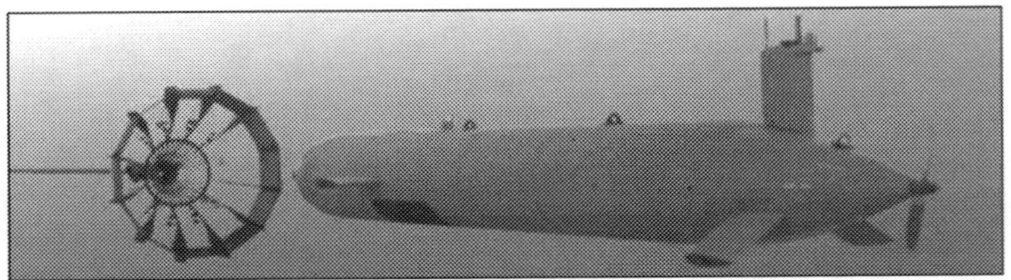

Figure 11-10. Towed docking system for autonomously docking the Explorer AUV
(Source: ISE Canada)[18]

- **Acoustic Positioning:** The recovery process utilizes acoustic homing signals to guide the AUV towards a taut line suspended from the USV.

- **Reduced Human Risk:** By automating the most dangerous aspects of the launch and recovery process, the system reduces the need for personnel to be exposed on deck during adverse conditions.

---

17 Sarda & Dhanak, 2018
18 ISE, 2025

- **Demonstrated Feasibility:** Modelling and simulation using tools like OrcaFlex have shown that the system can function effectively in low to moderate sea states, with ongoing research aimed at extending its operational envelope.

Figure 11-11. ARV-I AUV with docking station for wireless battery charging and transferring collected data
(Source: Boxfish Robotics)[19]

Custom and Modular LARS Solutions

Custom solutions are developed to meet specific mission requirements or to integrate with unusual platforms. On the other hand, modular systems are designed with interchangeable components that can be reconfigured to handle different vehicles or operational conditions. For example, a modular LARS might use interchangeable winch assemblies that can be swapped based on the weight and size of the vehicle. A trials set-up in harbour may be valuable for testing and training operators on LARS (Figure 11-12).

---

[19] https://www.boxfishrobotics.com/products/arv-i/arv-i-features/?utm_source=unmanne dsystemstechnology.com&utm_medium=referral

Figure 11-12 AUV Amogh being deployed from harbour
(Source: Larsen & Toubro)[20]

Some missions require customized solutions, such as deep-water recovery of heavy vehicles or recovery in highly dynamic environments. Systems like NOV-BLM's LARS for offshore energy applications exemplify the development of high-capacity, robust systems designed for extreme conditions.[21]

In military applications, LARS may be integrated with other mission systems—for example, using onboard sensors, acoustic homing, and guidance algorithms to enable rapid recovery during high-stress operations. These systems are often designed for stealth and rapid redeployment.

Design Approaches for LARS

The design of an effective LARS is a multidisciplinary challenge that involves mechanical, hydrodynamic, control, and materials engineering. The following subsections describe common design approaches and key considerations.

---

[20]   Siddiqui, 2022
[21]   NOV, 2025; IHC Offshore.

Mechanical Design Considerations

Structural Components

The primary structural components of a LARS include cranes, winches, A-frames, and telescopic arms. These components must be designed to withstand both static loads (the weight of the underwater vehicle) and dynamic loads (forces generated by waves, vessel motion, and rapid acceleration/ deceleration). Engineers often use finite element analysis (FEA) to simulate stress distribution, fatigue, and potential failure modes. Materials chosen for these components are typically high-strength, corrosion-resistant alloys or composites that can operate in saltwater environments. Manufacturing techniques such as precision machining, welding under controlled conditions, and advanced composite layup are employed to produce components that meet the stringent performance requirements of LARS.

Winch and Cable Systems

Winches are critical for controlling the descent and ascent of the vehicle. Modern winch systems incorporate variable-speed controls and are designed for rapid deployment and recovery while providing sufficient braking to manage dynamic loads. The cable used for ROV umbilicals, or for lowering AUVs, must balance flexibility with high tensile strength and resistance to corrosion and abrasion. Some systems also include tether management systems (TMS) to minimize drag and entanglement during operations.

Shock Absorption and Heave Compensation

Shock absorbers and heave compensation systems are designed to mitigate the effects of vessel motion on the vehicle during launch and recovery. Systems may use hydraulic dampers, spring-loaded mechanisms, or active control algorithms that adjust winch speeds in real time. In some cases, telescopic arms are integrated with shock-absorbing elements to smooth the transition through the turbulent splash zone.

Hydrodynamic Design

Hydrodynamic performance is critical in ensuring the smooth passage of the vehicle through the water. Designers must account for the following aspects:

- **Drag Reduction:** Minimizing hydrodynamic drag is essential for energy efficiency and for preventing undue strain on mechanical components. Streamlined shapes and the careful selection of materials can help reduce drag.

- **Flow-Induced Vibrations:** Vibration due to turbulent water flow can lead to fatigue and wear on both the LARS and the underwater vehicle. CFD simulations help engineers predict and mitigate these effects.

- **Alignment and Stability:** Ensuring that the vehicle remains aligned during deployment is crucial. This may involve designing guiding structures, such as rails or funnels that direct the vehicle into the correct orientation before it is fully in the water.

Control and Automation

Automation in LARS design is increasingly vital to ensure safe and efficient operations. Key aspects include:

- **Sensor Integration:** Modern LARS incorporates an array of sensors including accelerometers, gyroscopes, GPS, and acoustic position transducers. These sensors provide real-time feedback on vessel motion and vehicle position, allowing the control system to adjust operations dynamically.

- **Automated Control Algorithms:** Advanced control algorithms—often developed using simulation tools like MATLAB/Simulink or OrcaFlex—enable the system to compensate for external disturbances automatically. These algorithms are responsible for adjusting winch speeds, aligning the vehicle, and ensuring a smooth transition between air and water.

- **Human-Machine Interface (HMI):** Even in automated systems, a robust HMI is necessary for monitoring and manual override in emergency situations. Clear display of sensor data, load status, and system health are essential components of modern LARS control panels.

Integration with Vessel Systems

The LARS must be seamlessly integrated into the support vessel's structure and systems. This involves ensuring that the mounting points, power supplies, control interfaces, and communication networks are compatible. In many cases, LARS is designed as a modular system that can be retrofitted to existing vessels, minimizing the need for extensive vessel modifications. Dynamic positioning (DP) systems on the vessel also play a crucial role by maintaining the precise location required for safe launch and recovery operations.

Examples of LARS Solutions

Several customized LARS solutions have been developed to meet the specialized requirements of different applications. Some examples illustrate how different approaches are used in practice.

An integrated LARS mechanism with the containerised storage solution for HUGIN is shown in Figure 11-13.

Figure 11-13. Launch and recovery ramp for Hugin AUV integrated with its storage container (Source: Kongsberg Maritime)[22]

REMUS Launch and Recovery Systems

The REMUS Launch and Recovery System developed by the Woods Hole Oceanographic Institution (WHOI) has been used for hundreds of launches and recoveries in various sea states (up to sea state 5), proving its reliability. Key features include:

- **A-Frame with Integrated Heave Compensation:** This system is mounted on the stern of the support vessel and uses a tilting A-frame to gently lower the vehicle into the water (Figure 11-14). The cradle includes shock absorbers to dampen the effects of waves.

---

[22] Polar AUV Guide: Deployment and Recovery

- **Acoustic Homing and Guidance:** Sensors integrated into the system help align the vehicle during launch and recovery.

Figure 11-14. LARS for REMUS AUV
(Source: WHOI)[23]

NOV-BLM LARS

NOV-BLM has developed a launch and recovery system specifically designed for offshore energy applications, with demonstrated reliable performance in high sea states (up to 5).[24] Their system features:

- **High Lifting Capacity:** Configurations are available from 3 to 50 tons, making it suitable for large underwater vehicles and crew boats.

- **Fully Electric Operation:** The system is designed to work efficiently in harsh offshore environments with minimal maintenance.

- **Adjustable and Shock-Absorbent Cradle:** Ensures the vehicle is securely held and protected during the recovery phase.

Military and Salvage Applications

Military operations require robust and often stealthy LARS solutions for the rapid deployment of unmanned underwater systems. An example is provided by L3 Harris, which has demonstrated an integrated system for launching and recovering AUVs from an underway submarine.[25] Key features include:

---

[23] https://images.whoi.edu/view-item?i=372108&WINID=1751450597406
[24] IHC Offshore; NOV, 2025.
[25] L3 Harris, 2023

- **Torpedo Tube Launch and Recovery (TTL&R):** In this approach, a compact recovery mechanism is integrated into a submarine's torpedo tube, allowing an AUV to be launched and later recovered without exposing the host vessel.
- **Rapid Deployment:** The system was successfully demonstrated in tests where an AUV was deployed to perform reconnaissance missions while the submarine remained covert.
- **Operational Advantages:** Such systems enhance the capability of manned-unmanned teaming, allowing submarines to extend their surveillance and data-gathering operations without risk to crew.

Deep-Water Salvage Systems

Although primarily designed for salvage operations, systems like the Flyaway Deep Ocean Salvage System (FADOSS) share many design principles with LARS for underwater vehicles.[26] The system is designed to be airlifted and installed quickly on a vessel of opportunity, providing flexibility for emergency operations. Capable of lifting objects from several tons, FADOSS has been used to recover aircraft and other large equipment from extreme depths. A key component of the system is the Ship Motion Compensator, which uses a pressurized ram cylinder and sheaves to compensate for the vessel's motion during lifting operations.

Summary of LARS Solutions

Table 11-1 provides a summary of several representative LARS solutions, their capacities, example applications, and key features.

Testing and Trials: Overview

Underwater vehicles operate in environments where errors can have severe consequences, including loss of expensive equipment, damage to support vessels, or even threats to human life. Operational Test & Evaluation (OT&E) is necessary to determine the readiness of the overall system for deployments. However, OT&E of complex robotic systems such as AUVs is notoriously difficult and evolving. Hence, a framework for Experimental T&E (ET&E), merging research, development, acceptance and operational testing through designed experiments is appropriate.[27]

---

[26]   https://www.navy.mil/Resources/Photo-Gallery/igphoto/2002536255/

[27]   Keane & Joiner, 2020

Table 11-1. Summary of Solutions of LARS

LARS Solution	Lifting/ Deployment Capacity	Example Application	Key Features
WHOI REMUS LARS	Suitable for small AUVs (REMUS series)	Oceanographic research; AUV deployments in sea state up to 5	A-frame system; integrated heave compensation; acoustic guidance; proven reliability with hundreds of launches and recoveries
NOV-BLM's Offshore LARS	3 to 50 tons	Offshore energy operations; unmanned crew boats, small submarines	Fully electric; adjustable, shock-absorbent cradle; modular design; designed to operate in high sea states; robust for heavy loads
USV-Based Automated LARS	Suitable for lightweight AUVs (e.g., REMUS 100)	Autonomous deployment and recovery from USVs in moderate sea states	Automated winch and telescopic arm; acoustic positioning for precise recovery; minimizes human exposure; integration with unmanned surface vehicles (USVs)
L3Harris Torpedo Tube LARS	Compact system for AUVs; designed for submarines	Military reconnaissance; covert AUV deployment from submarines	Integrated within torpedo tubes; rapid and stealthy launch/recovery; enables manned-unmanned teaming; supports underwater reconnaissance while maintaining vessel stealth
Flyaway Deep Ocean Salvage System (FADOSS)	Up to 20+ tons	Salvage operations; recovery of aircraft, submersibles from extreme depths	Modular, airlift-deployable; ship motion compensator (SMC); specialized winch and storage reel; designed for rapid installation on vessels of opportunity; proven in deep water recoveries

Significance

Thorough testing minimizes these risks by ensuring that every subsystem—from propulsion and navigation to communication and power management—functions correctly under a range of conditions. The incremental approach to testing allows designers and engineers to validate assumptions, refine control algorithms, and optimize mechanical designs before exposing the vehicle to the full rigors of open water. AUVs are complex systems and not necessarily suited to the mould of a traditional T&E Master Plan (TEMP). T&E must be flexible and opportunistic, while comprehensive enough to extract quantifiable measures of capability from a probabilistic system.[28]

---

[28]   Macias, 2008

Performance Verification and Optimization

Testing provides quantitative data on vehicle performance, including parameters such as speed, endurance, manoeuvrability, depth rating, and sensor accuracy. This data is critical for calibrating the vehicle's control systems and ensuring that operational specifications are met. Moreover, trial phases help in identifying performance bottlenecks and areas for improvement, which can then be addressed in subsequent design iterations.

Integration and Interface Testing

Underwater vehicles are highly complex systems, integrating hardware, software, and communication networks. Testing ensures that these components work seamlessly together. Hardware-in-loop (HIL) testing, for instance, simulates the dynamic interactions between the vehicle and its environment, enabling engineers to verify the integration of control systems with physical hardware. Such integration tests are fundamental to achieving reliable operation in the real-world marine environment.

Regulatory and Certification Requirements

For many applications—especially in commercial and military contexts—underwater vehicles must meet rigorous regulatory standards and certification requirements. Systematic testing provides the evidence needed to obtain certifications and ensures compliance with safety and performance standards.

Phases of Underwater Vehicle Testing and Trials

Underwater vehicle trials typically follow a phased approach, each with specific objectives and conditions:

- **Hardware-in-Loop (HIL) Testing:** Performed in the laboratory or simulation environment, HIL testing allows developers to integrate and test real hardware with simulated software and environmental inputs.

- **Harbour Trials:** Conducted in controlled, calm waters (such as a harbour or a designated testing basin), these trials serve as an intermediate step between laboratory tests and open-sea operations.

- **Sea Trials:** The final phase of testing, sea trials, exposes the vehicle to real ocean conditions, including waves, currents, and variable weather, providing a comprehensive assessment of performance in its intended operational environment.

Each phase is designed to gradually increase the level of complexity and environmental realism, thereby reducing risk and providing incremental validation of system performance. These phases are elaborated further below.

Hardware-in-Loop (HIL) Testing

In Hardware-in-loop (HIL) testing, the actual hardware components of the underwater vehicle (such as controllers, sensors, and actuators) are interfaced with a simulated environment. The simulation could mimic the dynamic behaviour of the ocean, including hydrodynamic forces, vessel motion, and sensor feedback. This approach allows engineers to test the real-time performance of the control system without risking the actual vehicle in the field.

Simulation environments allow for rapid testing of various scenarios, enabling engineers to adjust control parameters, refine algorithms, and evaluate system responses under a wide range of conditions.[29] HIL testing is crucial for verifying that the real hardware components interact correctly with the vehicle's software. This is particularly important for ensuring that sensor data is accurately interpreted and that actuator commands produce the expected results.

Implementation of HIL Testing

In a typical HIL setup for underwater vehicles, the following components are integrated:

- **Control Hardware:** This includes the vehicle's central processing unit, motor controllers, and sensor interfaces.
- **Simulated Environment:** Using software such as MATLAB/Simulink, OrcaFlex, or specialized marine simulation tools, the dynamic behaviour of the underwater environment is modelled.
- **Real-Time Data Exchange:** Communication between the physical hardware and the simulation occurs in real time. The hardware receives simulated sensor inputs (e.g., simulated depth, speed, and orientation) and sends actuator commands back to the simulation.
- **Feedback and Monitoring:** Engineers use data acquisition systems and real-time monitoring dashboards to observe system behaviour,

---

[29]   Dynautics, 2021

analyze performance, and identify any discrepancies between expected and actual responses.

Challenges in HIL Testing

Despite its many benefits, HIL testing poses several challenges. For example, a research team developing an AUV encountered issues during HIL testing when the simulated water currents did not accurately reflect the transient turbulence seen in real sea conditions. This led to discrepancies in the vehicle's steering response. Refining the simulation model and incorporating higher-resolution turbulence data was necessary to resolve the issue.

Other challenges that could be encountered are listed below:

- **Simulation Accuracy:** The fidelity of the simulation directly impacts the effectiveness of HIL testing. Inaccurate models of hydrodynamic forces or sensor behaviour can lead to erroneous conclusions.

- **Integration Issues:** Ensuring seamless communication between the hardware and simulation software can be complex, particularly when dealing with high-speed data streams or multiple sensor inputs.

- **Real-Time Constraints:** HIL testing must operate in real time, which requires that the simulation software can compute complex environmental dynamics without lag. Latency issues can compromise the validity of test results.

- **Hardware Limitations:** Sometimes, the physical hardware used in HIL testing may not fully represent the final deployed configuration, leading to gaps in testing coverage.

Harbour Trials

Harbour trials are conducted in a controlled, sheltered environment such as a harbour, dock, or testing basin. They serve as an intermediate step between laboratory HIL tests and open-sea trials. In harbour trials, the underwater vehicle is deployed in sheltered water, allowing engineers to assess its performance under relatively calm and predictable conditions. Key objectives of harbour trials are as follows:

- **Verification of System Integration:** Harbour trials validate the integration of the underwater vehicle with its launch, recovery, and support systems.

- **Initial Performance Assessment:** Parameters such as manoeuvrability, speed, sensor accuracy, and control system responsiveness are measured.
- **Safety Testing:** Operators can test emergency procedures, such as abort sequences and manual overrides, in a safe environment.
- **Calibration and Tuning:** Fine-tuning of control systems, sensor calibration, and adjustments to mechanical systems are performed based on observed behaviour.
- **Crew Training:** Harbour trials provide an opportunity for training operators and crew, enabling them to become familiar with the system before full-scale sea trials.

Conducting Harbour Trials

Harbour trials are typically planned and executed in several stages:

- **Static Tests:** The vehicle is deployed in a relatively stationary state to test basic functionalities such as power systems, sensor readings, and initial control responses.
- **Low-Speed Manoeuvring:** The vehicle is then operated at low speeds to evaluate steering, navigation, and response to commands. This stage often includes tethered tests and short range dives.
- **Integration with Support Systems:** The vehicle is tested with its associated launch and recovery systems, which may include A-frames, cranes, or winches. These tests focus on the mechanical and electrical interfaces between the vehicle and the support vessel.
- **Emergency Drills:** Simulated emergency scenarios (e.g., loss of communication, sudden sensor failure) are executed to verify that safety protocols function correctly.

Challenges during Harbour Trials

Although harbour trials are conducted in relatively calm conditions, they still present a range of challenges. During harbour trials for a new AUV design, the team may discover that the vehicle's buoyancy control system is sensitive to slight temperature changes in the harbour water. This may result in erratic depth control during low-speed manoeuvres. The issue can be addressed by recalibrating the buoyancy system and improving the insulation of critical components. Other challenges in harbour trials could be as follows:

- **Limited Environmental Variability:** Harbour conditions are generally calmer than open-sea environments. This can mask issues that might arise in more turbulent conditions.

- **Space Constraints:** Harbours have limited space, which can constrain the manoeuvring of larger vehicles or affect the performance of dynamic positioning systems.

- **Integration Complexity:** The transition from laboratory HIL testing to real-water conditions can reveal unforeseen integration issues—such as improper sealing of electrical components, buoyancy mismatches, or communication difficulties between the vehicle and support systems.

- **Operator Errors:** Harbour trials often serve as the first real-world test of the operator's ability to manage the vehicle. Human errors during deployment, retrieval, or emergency procedures can lead to minor incidents that need to be corrected.

Figure 11-15. Indian HE-AUV being deployed for lake trials
(Source: DRDO)[30]

Sea Trials

Sea trials are the final phase of testing, where the underwater vehicle is deployed in open-water environments that closely resemble its intended operational

---

[30] Menon, 2025

conditions. These trials are conducted at sea and expose the vehicle to the full range of environmental challenges, including waves, currents, variable weather and complex underwater terrain.[31] Key objectives of sea trials are as follows:

- Operational Performance Evaluation: Sea trials assess the vehicle's overall performance in terms of speed, endurance, manoeuvrability, depth capability, and sensor functionality under real-world conditions.

- Environmental Resilience: These trials test the vehicle's ability to withstand and operate in adverse weather conditions, high waves, strong currents, and other unpredictable ocean dynamics.

- System Reliability and Robustness: Continuous operation in the open sea helps identify potential reliability issues—such as component wear, unexpected interactions between subsystems, or degradation of sensor performance.

- Safety and Emergency Procedures: Sea trials validate the full range of safety protocols, including abort sequences, emergency recovery, and fail-safe mechanisms. Real-world scenarios such as loss of power or communication are simulated to assess the robustness of these procedures.

- Data Collection and Mission Simulation: For research and commercial applications, sea trials are critical for collecting operational data. This data is used to refine operational parameters, improve mission planning, and optimize sensor performance.

Execution of Sea Trials

Sea trials are usually conducted in a series of stages:

- **Shakedown Runs:** Initial short-duration deployments are conducted to verify that the vehicle can leave the support vessel, execute basic manoeuvres, and return safely. These runs often focus on verifying the functionality of critical systems such as propulsion, control, and communications.

- **Incremental Mission Profiles:** Once the shakedown phase is complete, the vehicle is tested under increasingly challenging conditions. This may involve deeper dives, extended missions, and more complex manoeuvres such as obstacle avoidance and automated path following.

---

[31] Xiang Liu, 2025; Zhang, 2020.

- **Full-Mission Simulations:** The final phase of sea trials replicates the full operational mission, including long-duration deployments, data collection, and interaction with other systems (such as docking stations or unmanned surface vehicles).

- **Post-Trial Analysis:** Data collected during sea trials is analyzed to assess performance, identify failures, and recommend modifications. This analysis is crucial for certification, regulatory approval, and iterative design improvements.

Challenges during Sea Trials

Sea trials present the most rigorous test conditions and thus reveal many challenges that may not be apparent during HIL or harbour trials:

- **Environmental Uncertainty:** Open-sea conditions are unpredictable. Sudden weather changes, unexpected currents, and large waves can cause the vehicle to behave erratically or even damage sensitive components.

- **Integration of Systems:** The complexity of integrating mechanical, electronic and software systems becomes more apparent in a real-world setting. For example, the interaction between dynamic positioning systems on the support vessel and the vehicle's control system can be challenging to synchronize.

- **Communication Disruptions:** Underwater communication remains one of the most challenging aspects of sea trials. Acoustic modems may suffer from multi-path interference, low bandwidth, or dropouts due to environmental noise. This can lead to delays in command execution and reduced situational awareness.

- **Operational Fatigue:** Extended missions can reveal issues such as battery degradation, thermal management challenges, or gradual drift in sensor calibration. These factors require robust design and redundancy.

- **Safety Incidents:** Despite extensive pre-trial testing, unexpected incidents can occur during sea trials. For instance, a vehicle might encounter an unanticipated obstacle or experience a temporary loss of control. Such incidents, while undesirable, provide invaluable data for improving system resilience and emergency procedures.

Typical Issues during Testing Process

Testing underwater vehicles is a complex, iterative process. The key challenges likely in each stage of testing have been mentioned above. Some common issues likely to be faced across various stages of the testing process of UUVs are listed below for ready reference.

Simulation vs. Reality Discrepancies

One of the most common challenges in HIL testing is the gap between simulated conditions and real-world dynamics. For example, a simulation may not accurately capture the chaotic behaviour of ocean currents or the rapid fluctuations in sea state. This can lead to control algorithms that perform well in the lab but fail under actual conditions. Continuous refinement of simulation models using data from harbour and sea trials is necessary to minimise these discrepancies.

Sensor and Communication Issues

Underwater vehicles rely heavily on sensors for navigation, control, and data collection. During testing, issues such as sensor drift, calibration errors, and data latency can significantly impact performance. Acoustic communication, which is often used in underwater trials, is particularly prone to interference from ambient noise and multipath effects. Such issues can lead to delays in command execution or even temporary loss of communication, complicating the testing process.

Mechanical and Structural Failures

The marine environment is inherently harsh, with saltwater, pressure, and dynamic loads all contributing to wear and tear on mechanical components. During sea trials, even minor design flaws in the vehicle's structure or in the launch and recovery system can lead to failures. Problems such as cable fatigue, hydraulic system leaks, and joint wear are not uncommon, and they often require on-the-fly modifications and post-trial repairs.

Integration and Interface Mismatches

Underwater vehicles are composed of many interdependent subsystems. Ensuring that these subsystems interact seamlessly is a major challenge. Integration issues may arise from incompatible communication protocols, mismatched electrical interfaces, or differing response times between subsystems. Such problems often become apparent only during integrated testing in harbour or sea trials.

Environmental and Operational Uncertainties

The variability of ocean conditions is one of the most unpredictable factors affecting testing. Sudden changes in weather, unexpected tidal currents, or the presence of debris can impact the performance of an underwater vehicle during sea trials. Operators must be prepared to handle these uncertainties, often requiring backup procedures and contingency plans.

Human Factors and Training

Despite the increasing automation of underwater vehicles, human operators still play a critical role in managing trials. Operator errors, whether due to insufficient training or misinterpretation of sensor data, can lead to incidents during testing. Continuous training, simulation exercises, and the development of robust human-machine interfaces are essential to mitigate these risks.

Planning a Sequence of Tests and Trials

An idealised sequence of activities in underwater vehicle trials involves a phased approach starting from controlled simulation environments and progressing to full-scale sea trials.

Each AUV mission profile has unique requirements with distinct operational priorities; however, a number of underlying capability factors or measures of effectiveness have been identified (e.g., navigational accuracy, sensor effectiveness, and clearance rates). These universal factors form the basis of most AUV capabilities, such that Experimental Test & Evaluation (ET&E) can be standardised to an extent, thereby enabling system comparison to determine the suitability of different AUVs for various operating profiles.[32]

The main goals of T&E are to reduce risk and provide information for acquisition decisions or answer capability questions posed by the end-user. The two key phases in constructing the framework are the Design of Experiments (DoE) approach to establishing key factors and outputs, and development of a test plan.[33] Once combined, such a framework can test the technical and operational measurements required to determine whether the

---

[32] Kass, 2015
[33] Antony, 2014

AUV capability achieves its intended purpose, and to answer typical questions posed by decision-makers (potential customers or funding agencies).[34]

Deriving a Process Flow (PF) diagram outlines the typical operational flow for AUV deployments which is critical in understanding where technical and operational measurements are appropriate. This can form the backbone of the ET&E framework (Figure 11-16).

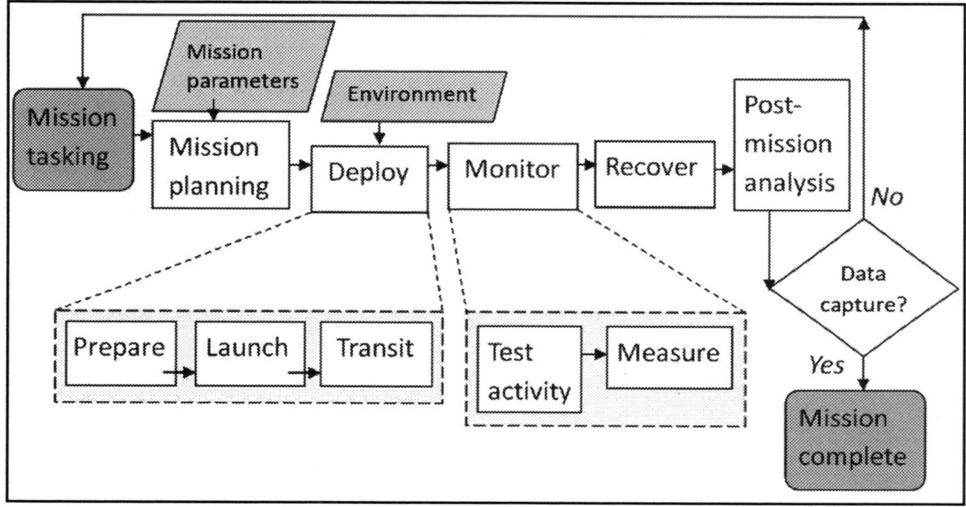

Figure 11-16. Process flow diagram for AUV Operational Test & Evaluation[35]

Example of MCM AUV

Consider an AUV developed for Mine Countermeasures (MCM) role involving search. The AUV can be considered as part of systems-of-systems for MCM effect. For example, in the context of a most basic MCM Critical Operating Issue (COI), 'Can the AUV capability search and detect a mine-like object?', the requirements of the vehicle can be developed by decomposing Measures of Effectiveness (MOE) and Measures of Performance (MOP) for many systems off-board the AUV and over longer-term MCM mission timeframes.[36]

---

[34] Gooding, 2001

[35] Keane & Joiner, 2020

[36] Gooding, 2001

Four core activities were derived for this Mine Countermeasures scenario. These are detailed in Table 11-2, and necessary test results can be derived from variants of these activities.

Table 11-2. Purpose and overview of test activities for an MCM AUV[37]

Activity	*Purpose of Test*	*Overview*
Payload	Effectiveness of sonar and camera payloads	Payload effectiveness is tested over a dedicated range of targets. These targets could be mine-like objects or unclassified targets such as crayfish pots. Effectiveness is measured as a probability of detection. As such, numerous repeat runs are required.
Communications	Range and reliability of different communications methods	This is a continuation of range and reliability tests. Building a database of communication performance during ET&E is useful as it contributes to further development for collaborative autonomy developments
Vehicle Dynamics	Characterise AUV dynamics through different manoeuvres when subject to environmental factors.	Vehicle dynamics during AUV missions is a critical factor in the quality of payload data.
MCM Search	Characterise AUV ability to detect Mine Like Objects in a range of operating environments.	The MCM search activity assesses the AUV capability once sufficient results from other activities have been derived. This activity aims to measure tactical level outputs such as clearance rate, probability of detection, coverage achieved, and mission completion rates.

The user effectiveness (COI) might be evaluated as the answer to a question such as, "Can the AUV detect a Mine-Like Object?". To answer this, we need to ask further questions to properly understand the capability, such as sensor accuracy and confidence of classification. The payload effectiveness, communications, and vehicle dynamics activities are necessary to build the foundational understanding of AUV capability before determining the effectiveness of the COI. This order is due to the relationship between navigational accuracy, sensor effectiveness, and vehicle dynamics. The navigation system is a deliberate omission from this list as navigation accuracy is typically characterised sufficiently during other phases of T&E. Thus for ET&E, navigation is measured as a subset of payload performance.

---

[37]   Keane & Joiner, 2020

Test & Evaluation Plan

The 'Design of Experiments' methodology and first-principles approach discussed above can be used to draw up an Experimental Test & Evaluation (ET&E) framework that combines the process flow (PF) and operational requirements. An AUV ET&E framework (Keane & Joiner, 2020) is shown in Figure 11-17.

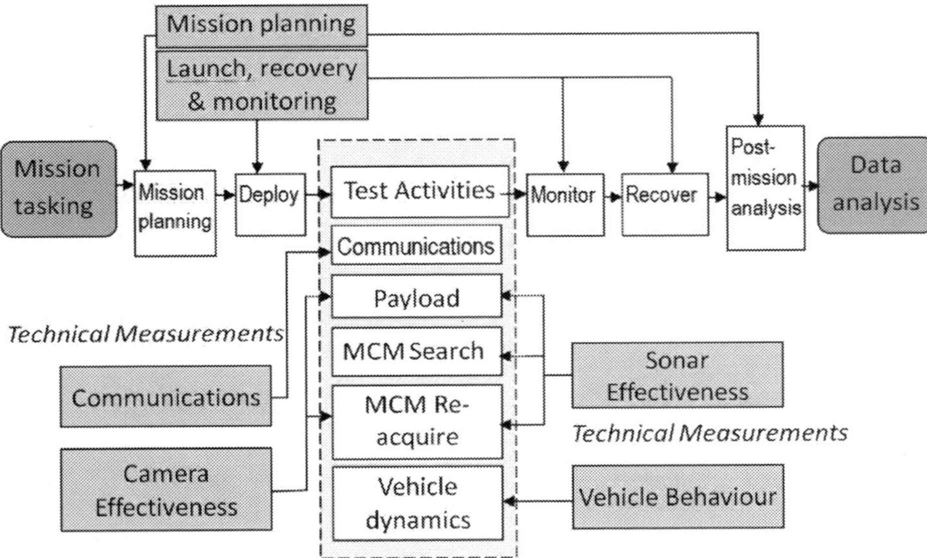

Figure 11-17. Experimental Test & Evaluation Framework for an AUV[38]

Table 11-3 shows a typical sequence of trials with approximate durations. Actual durations may vary based on the vehicle's complexity, environmental conditions, and specific mission requirements. The durations indicated (total about 20 to 30 weeks) are idealised estimates for a medium-complexity underwater vehicle.

---

[38]   Keane & Joiner, 2020

Table 11-3. Idealised example for sequence of AUV testing and trials

Phase	*Activity Description*	*Approx. Duration*
1. Hardware-in-Loop (HIL) Testing	• Integration of control hardware with simulation environment. • Testing of sensor inputs, control algorithms, and actuator responses. • Iterative refinement based on simulation data.	4–6 weeks (as per complexity and iterations)
2. Laboratory Component Testing	• Bench testing of individual subsystems (e.g., propulsion, buoyancy, communication). • Stress and fatigue testing of critical components.	2–4 weeks
3. Controlled Environment (Harbour) Trials – Preliminary	• Static deployment in a harbour or testing basin. • Verification of basic deployment and recovery procedures. • Initial operator training and system calibration.	1–2 weeks
4. Controlled Environment (Harbour) Trials – Dynamic	• Low-speed manoeuvring and integration tests with launch/recovery systems. • Simulation of emergency procedures and manual override tests.	2–3 weeks
5. Incremental Sea Trials – Shakedown	• Short-duration sea trials near the support vessel. • Verification of system performance under real water conditions. • Collection of baseline performance data.	1–2 weeks
6. Incremental Sea Trials – Extended Missions	• Progressive increase in mission complexity (deeper dives, longer durations, complex manoeuvres). • Testing of full mission profiles and emergency procedures under variable conditions.	3–4 weeks
7. Full-Scale Sea Trials and Final Certification	• Comprehensive deployment in operational sea conditions. • Performance evaluation over extended missions (full-endurance, high-speed manoeuvres, deep dives). • Final data collection for certification and regulatory compliance.	4–6 weeks
8. Post-Trial Analysis and Feedback	• Detailed review and analysis of all trial data. • Identification of design improvements and operational adjustments. • Documentation and training updates.	2–3 weeks (overlapping with trial phases)

Conclusion

We have seen that deployment and recovery of UUVs is a challenging part of operations, for which reliable LARS and their proper operation are essential. The types of LARS solutions and considerations in their selection and design have been discussed. Operator training and frequent operations are essential to gain proficiency in handling launch and recovery at sea under variable weather conditions. Along with safety of the personnel involved in handling operations, safety of the expensive UUV representing years of effort is also essential. Standardised solutions for launch and recovery are available for established models of AUVs. In case of new development AUV, design of the LARS should be given due consideration during the vehicle design process itself.

Underwater vehicle testing and trials are an indispensable part of the development process, ensuring that vehicles can operate reliably in the harsh and unpredictable marine environment. The testing process typically progresses from hardware-in-loop (HIL) testing in the laboratory, through controlled harbour trials, and ultimately to full-scale sea trials. As discussed here, significant introspection is necessary about the AUV's purpose and mission in order to plan the scope, purpose and sequence of trials.

Each phase serves a unique purpose, from early risk mitigation and subsystem integration, to comprehensive operational performance evaluation. This progression reflects the strategy of gradually increasing complexity and exposure to risk. Each phase builds upon the successes and lessons of the previous stage; hence any short cuts may be catastrophic.

References

Antony, J. (2014). *Design of Experiments for Engineers and Scientists*. London: Elsevier.

Ballard, R.D. (2021). *Into the Deep: A Memoir from the Man who Found Titanic*. National Geographic.

Butler, B. (2018). *Into the Labyrinth: The Making of a Modern-Day Theseus*. Canada.

Christ, R. S. (2011). *The ROV Manual: A User Guide for Observation Class Remotely Operated Vehicles*. Netherlands: Elsevier Science.

Gooding, T. R. (2001). *A framework for evaluating advanced search concepts for Multiple Autonomous Underwater Vehicle (AUV) Mine Countermeasures (MCM)*. Masters Thesis, Massachusetts Institute of Technology, USA. Retrieved December 2023 from https://www.semanticscholar.org/paper/A-framework-for-evaluating-advanced-search-conceptsGooding/7be3b3f3b5c3a6c7bac73fa5b981a33de8096174

Griffiths, G., Millard, N., & Rogers, R. (2003). Logistics, Risks and Procedures concerning Autonomous Underwater Vehicles. In *Technology and Applications of Autonomous Underwater Vehicles* (pp. 279-293). CRC Press. doi:10.1201/9780203522301.ch16

Hydro International (2020, January 16). *This Is How an Unmanned Launch and Recovery System Operates*. Retrieved from Hydro International: https://www.hydro-international.com/content/news/this-is-how-an-unmanned-launch-and-recovery-system-operates

ISE (2025). *Autonomous Docking: Discover Unlimited Range: Surfacing Optional.* Retrieved from International Submarine Engineering Ltd.: https://ise.bc.ca/product/dock/

Kass, R. (2015). Twenty-one Parameters for Rigorous, Robust, and Realistic Operational Testing. *ITEA Journal,* 36, 121–138.

Keane, J., & Joiner, K. F. (2020). Experimental Test and Evaluation of Autonomous Underwater Vehicles. *Australian Journal of Multi-Disciplinary Engineering,* 16(6), 1-13. doi:10.1080/14488388.2020.1788228

L3 Harris (2023, July 03). *Successful Launch and Recovery of Autonomous Underwater Vehicle from Underway Submarine.* Retrieved from L3 Harris: https://www.l3harris.com/newsroom/editorial/2023/07/successful-launch-and-recovery-autonomous-underwater-vehicle-underway

IHC Offshore (n.d.). *Launch and recovery systems (LARS).* Retrieved from IHC Offshore Energy: https://www.royalihc.com/offshore-energy/offshore-equipment/launch-and-recovery-systems

Macias, F. (2008). The Test and Evaluation of Unmanned and Autonomous Systems. *ITEA Journal,* 29, 388–395.

Menon, A. K. (2025, April 20). *India's High Endurance AUV Program Undergoes Developmental Trials.* Retrieved from *Naval News*: https://www.navalnews.com/naval-news/2025/04/indias-high-endurance-auv-program-undergoes-developmental-trials/

Dynautics (2021, June 15). *Modelling, testing & sea trials – all without leaving the lab.* Retrieved from Dynautics: https://www.dynautics.com/modelling-testing-and-sea-trials-all-without-leaving-the-lab/

NOV (2025). *Launch and Recovery System (LARS).* Retrieved January 2025 from NOV.com: https://www.nov.com/products/launch-and-recovery-system-lars

Pantheeradiyil, A. (2024, October 7). *How ROVs and divers are transforming underwater inspections for safety and efficiency.* Retrieved from blueyerobotics.com: https://www.blueyerobotics.com/blog/how-rovs-and-divers-are-transforming-underwater-inspections-for-safety-and

Peruzzi, L. (2023, July 5). Fincantieri NexTech will take multirole SAND USV to REPMUS 2023 in Portugal. Retrieved from European Defence Review EDR online: https://www.edrmagazine.eu/fincantieri-nextech-will-take-sand-usv-to-repmus-2023-in-portugal

Polar AUV Guide: Deployment and Recovery. (n.d.). Retrieved May 2025 from Society for Underwater Technology: http://wsprdaemon.org/polar/deployment-and-recovery.html

Ray, A., Singh, S., & Seshadri, V. (2011). Underwater Gliders: Force Multipliers for Naval Applications. *Warship 2011: Naval Submarines and UUVs.* Bath, UK: Royal Institution of Naval Architects. doi:10.3940/rina.ws.2011.03

Rumson, A. G. (2021). The application of fully unmanned robotic systems for inspection of subsea pipelines. *Ocean Engineering,* 235. doi:https://doi.org/10.1016/j.oceaneng.2021.109214.

Sagar, M. M., Konara, M., Picard, N., & Park, K. (2025, February). State-of-the-Art Navigation Systems and Sensors for Unmanned Underwater Vehicles (UUVs). *Applied Mechanics,* 6(1). doi:https://doi.org/10.3390/applmech6010010

Sarda, E. I., & Dhanak, M. R. (2018, September). Launch and Recovery of an AUV from a Station-Keeping USV. *IEEE Journal of Oceanic Engineering.* doi:DOI: 10.1109/JOE.2018.2867988

Siddiqui, Huma (2022, December 9). *Indian Navy seeks indigenous underwater vehicles to optimise unmanned tech.* Retrieved September 2025 from *Financial Express*: https://www.financialexpress.com/business/defence-indian-navy-seeks-indigenous-underwater-aerial-vehicles-to-optimise-unmanned-tech-know-all-about-it-2907602/

Thales (2020, November 26), *Thales to deliver the World's First Fully Integrated Unmanned Mine Countermeasures System for the Royal Navy and French Marine Nationale.* Retrieved from thalesgroup.com: https://www.thalesgroup.com/en/group/journalist/ press_release/ thales-deliver-worlds-first-fully-integrated-unmanned-mine

WHOI. (n.d.). *Launch and Recovery Systems (LARS).* Retrieved January 2025, from WHOI: https://www.whoi.edu/what-we-do/explore/underwater-vehicles/auvs/remus/launch-and-recovery-systems-lars/

Xiang Liu, S. X. (2025). Innovative Strategy and Practice of Using Underwater Robot for Marine Cable Inspection and Operation and Maintenance. *Cognitive Robotics.* Retrieved from https://doi.org/10.1016/j.rineng.2021.100201

Zhang, H. (2020). Subsea pipeline leak inspection by autonomous underwater vehicle. *Applied Ocean Research.* doi:10.1016/j.apor.2020.102321

12

Diving to the Future

Submarines have fascinated me since the beginning of my academy/college days about three decades ago. But it was a decade later, during post-graduation around 2005, that I encountered the world of unmanned underwater systems. Thanks to academic institutions (like IIT Bombay and IIT Madras) and scientific institutions (like NIOT Chennai and NIO Goa), batch after batch of engineering graduates have got enthused by this field. The ubiquity of the internet, increase in processing power, and enabling tech like Arduino, Raspberry Pi and Blue Robotics thrusters brought marine robotics to everyone's doorstep.

Enabling the Dive

Arduino and Raspberry Pi are widely recognized as path breakers in making robotics cheap and accessible for hobbyists, students, educators and professionals. First launched in 2005, Arduino was aimed specifically to enable non-engineers to easily build interactive devices, dramatically lowering the barrier to entry for embedded hardware projects. Raspberry Pi, launched in 2012, pursued a parallel mission by creating an affordable, compact single-board computer designed to teach programming and electronics, but quickly found adoption for a wide range of robotics and automation applications due to its versatility and low cost.[1] Both platforms sparked global communities and ecosystems of builders, libraries, and hardware add-ons. Their open-source approach and exceptionally low pricing made hands-on robotics possible outside traditional research labs or industry.

---

[1] Mathe et al, 2024

Along with open-source electronics, the availability of affordable, compact and modular components such as thrusters and sensors boosted this field tremendously. Blue Robotics was founded in 2014 in Torrance, California, with a goal to provide affordable, high-performance components for marine robotics.[2] Their components, like thrusters and watertight enclosures, have become the foundation for thousands of marine robotics projects globally. At major student competitions (such as RoboSub), over 70% of the teams are reported to use Blue Robotics parts.

An organisation that has played one of the most significant and influential roles in promoting marine and underwater robotics worldwide, is the Marine Technology Society (MTS). MTS actively supports technological development, collaboration, education, professional conferences, publications, and outreach activities related to ocean sciences and marine technology, including underwater robotics. Its journals and flagship events frequently highlight the latest advances in underwater robots, sensors, and autonomous systems for ocean exploration, industry, and conservation.[3] Marine Advanced Technology Education (MATE) Center is known for running the globally influential MATE ROV Competition.

These developments have democratised access to robust marine robotic technology and dramatically lowered barriers for students, researchers, hobbyists, and startups to build and deploy AUVs. This has fostered increased innovation and expanding accessibility in marine robotics technology.

Robotics Community

The National Institute of Oceanography is a CSIR lab perched on the hill of Dona Paula, Goa. Their 'Marine Robotics School' is a week-long workshop first organized by NIO (Marine Instrumentation Division) in 2012. My experience of attending the sixth edition of this event in November 2023 opened up a world of access to international experts and enthusiastic participants from academia, research, and industry. These lectures, demonstrations, and collaborative activities in marine robotics gave my small team of designers new impetus in our efforts. The UUV tech that I learnt from the Marine Robotics School forms the greater part of what I have tried to cover in this book.

---

2   https://bluerobotics.com/about/

3   https://www.marinebiodiversity.ca/transforming-ocean-science-the-impact-of-marine-technology-society-journals/

Most of the discussion in these chapters revolves around development, since the technology still appears as 'emerging'. However, the induction of AUVs in large numbers by several navies and research organisations worldwide is testimony to the maturity and variety of solutions already available. In Canada, British Columbia has been a hotbed of underwater technology development[4] and companies like International Submarine Engineering have been in this business since 1974.

It is only in the past five years that I have become aware of the 'deep tech' ecosystem in India. Before that, I wouldn't have been able to give a name to the diverse collection of entrepreneurs who are excited by cutting-edge technology. Our small design team for underwater robotics reached out and discovered connections, links, capabilities, and enthusiastic individuals in every corner of the country. We explored new ways of collaborations, funding, and prototyping. Some went away frustrated by the waiting, by our inability to follow up on our assurances, and some disappointed us as well. But with many of them, we were able to partner and participate in developing exciting, accessible, dual-use underwater tech.

Technology Trends

Integrating deep learning systems into mobile platforms will enhance robotic capabilities for navigation within dull, dirty, dangerous or dire situations. Key focus areas for developing AUV/UUV/ASV technologies for coordinated operation include the applications for port infrastructures and inspections of bridges, dams etc. on inland water bodies. Other focus areas involve the development of cloud-based service-oriented architectures, low-cost underwater acoustic modems, and long-range wireless devices. Additionally, the use of low-cost underwater vehicles and collaborative fleets can provide long-term surveillance and wide survey area capabilities.

The seamless coordination of autonomous vehicles, such as USVs, UUVs, and UAVs, is crucial for future autonomous marine missions, including autonomous recharging and high-throughput data transfer. Facilities needed for developing these technologies include an integrated underwater and above water network, sensors deployed in the port, and a shore station for interaction with the vessels. Solutions for fast and ultra-wide area bathymetric survey without a support vessel would involve multiple AUVs, an ASV, and a ground mission control station.

---

[4]   Jensen, 2022

Underwater Incidents

Underwater infrastructure could hardly be more critical for both our data and our energy needs. Over 95 percent of global data traffic passes through over 500 undersea fibre optic cables with a total length of roughly 1.4 million kilometres. Critical undersea infrastructure such as data or power cables are an attractive target for "gray zone operations" below the threshold of armed conflict.[5]

Rear Admiral Christian Meyer of the Germany Navy in a statement in June 2024 to the German parliament, stated that it is relatively easy "to do damage without getting caught." The vast expanse of the seas makes monitoring difficult. And even if vessels were to be caught red-handed, much critical infrastructure is located outside territorial waters where international law provides fewer options to go after suspect ships.[6]

In September 2024, the US military explicitly warned of Russian sabotage activities against undersea cables. Russia has invested heavily in the capabilities of the Russian military's 'Main Directorate for Deep-Sea Research' (GUGI) and operates a large fleet of ships, submarines, and underwater drones that could carry out attacks against critical infrastructure on the seabed.[7]

In October 2023, Finnish authorities suspected the Chinese cargo ship *Newnew Polar Bear* of damaging two submarine data cables and the Baltic-connector natural gas pipeline. During a stopover in the Russian port of Arkhangelsk, the ship was sighted with an anchor missing. The evidence suggested that critical infrastructure on the seabed had been damaged by the ship dragging its anchor. In August 2024, the Chinese authorities announced that the damage was indeed caused by the *Newnew Polar Bear*, but that it was an accident due to severe weather conditions.

In November 2024, another Chinese vessel, the *Yi Peng 3*, was suspected of having cut two submarine cables by dragging its anchor, one between Sweden and Lithuania and the other between Germany and Finland.

We could envisage the control and protection of the seabed by a network of active and passive sensors, managed by a network of nodes that could store and transfer data to centres on land. Drones (that could be dual-use) could

---

[5]   CCDCOE, 2019

[6]   Benner, 2025

[7]   Sciutto, 2024

be stationed on the seabed, with stations for recharging batteries, data transfer and remote maintenance activities. All these elements could be managed through an architecture of which submarines are also an integral part. Together, this system would ensure 'underwater situational awareness'.

The majority of the approximately 100 incidents of damage to undersea data cables that occur each year are not the result of sabotage, but of accidents and natural disasters.[8] Their inspection, repairs, maintenance and testing represent challenges of underwater work for which UUVs are essential.

Not-so-wild Ideas

'Nezha-SeaDart' is a Hybrid Aerial Underwater Vehicle (HAUV) that can not only move below water surface but also propel itself into the air and transform into an UAV. It was reported that the drone has completed 10-day trial at Qiandao Lake in Zhejiang province in China, proving its functionality.[9]

Recoverable UUVs can be used as offboard sensors and data harvesters from long-term seabed nodes. This will enable missions currently not possible for crewed submarines. Covert multi-domain sensors can provide coverage in places not previously reachable. Data harvesting and covert underwater communications could be used to get that data to users for planning and exploitations.

Torpedo-Tube Launch & Recovery (TTL&R) capability is integral to bringing UUVs into more regular current naval operational use. TTL&R for UUVs will help submarines contribute to subsea and seabed warfare operations. In late 2023, an HII REMUS 600 UUV was launched from the USS *Delaware* (a *Virginia* class SSN) torpedo tube.[10] The first-ever launch and recovery of an AUV from a torpedo tube for a tactical mission was achieved from the same submarine in June 2025.[11]

In July 2024, two firms (BMT and Ocius) announced a collaboration to provide a persistent and networked system of remote sensors. They aim to deploy a fleet of 1000 autonomous vessels globally within the next 10 years. It aims to develop and demonstrate the effectiveness of the 'Bluebottle'

---

[8]   CCDCOE, 2019

[9]   Kajal, 2024

[10]   https://nationalinterest.org/blog/buzz/us-navy-has-been-testing-torpedo-tube-launched-drone-207961

[11]   Mishra, 2025

uncrewed surface vessel (USV) for providing DaaS (Data as a service). They aim to co-develop a commercial data collection and management model, initially targeting environmental monitoring for offshore wind development projects.[12] Such initiatives could revolutionise maritime data collection and management.

Maturity for Defence

A report published by the RAND Corporation in 2019 aims to establish the state of the art in the development of autonomous systems and map how these might be effective in fighting wars. Salient findings from their analysis are as follows:[13]

- The capability for autonomous systems to interpret context and make independent decisions, particularly in a dynamic environment, is not realistic in the short term.

- The development of other capabilities, such as power generation and storage, might be limiting factors in developing greater autonomy.

- Use the unique features of autonomy to enable new CONOPS. Employing numerous systems carrying different kinds of sensors will require development of a system capable of fusing multiple inputs to create a common operational picture.

- Re-evaluate future fleet design architecture requirements in light of the state of the art of autonomous technology. However, some features of the proposed UUV architectures and networks may be more aspirational than likely.

- Autonomous systems will eventually need to make engagement decisions if those decisions are to be effective and timely.

- Develop a mechanism that allows humans to periodically check that an autonomous system is not misinterpreting its environment.

- Critically evaluate the viability of complex multi-mission platforms, and consider emphasizing simple but cooperating platforms.

---

[12]   BMT, 2024
[13]   Martin, Tarraf, et al., 2019

Reality versus Imagination

The effectiveness of unmanned or autonomous systems in meeting mission objectives can be truly evaluated only after the basic challenges of their deployment, reliable operation and communication are addressed. Most of the published data is naturally focused on vehicle and sensor characteristics, and operational data is limited to endurance, range and navigational accuracy.

To make a difference in the way scientific, industrial and military missions are conducted, the vehicles need to be reliable, easily deployable and extensively utilised. This poses a conundrum because, on the one hand, reliable AUVs are very costly and therefore too few to experiment with, and on the other, CONOPS cannot be developed merely by imagination or hearsay.

Vice Admiral Rob Gaucher, the US Navy's Commander Submarine Forces, told the Naval Leaders' Combined Naval Event 2024 conference May 24: "We're going to push to more exercises and development, to be able to realise our goal to do more and more uncrewed [activities]. We're ready to get the curve bending, to drive exponential growth in how often we're operating our uncrewed systems. Sure, there is a chance things don't go right, but we need to take the training wheels off and learn faster, particularly when we can shift risk from people to robots."[14]

Undoubtedly, underwater robotics is the future of maritime surveillance and numerous other dull, dirty, or dangerous activities at sea. However, as a researcher points out, "The technology exists; however, it is a lack of imagination that is disallowing its usage".[15]

It has been a privilege to witness energetic teams of engineers across India deploying their creations into the underwater domain, literally getting their feet wet and hands dirty, and also furiously tapping away at laptop keyboards ashore. That is what gives me confidence in the potential of underwater robotics.

In the tussle between reality versus imagination, the outcome would probably be somewhere in between. So the farther we can imagine, the lesser can reality hold us back.

---

[14] Willett, 2024

[15] Agarwala, 2022

References

Agarwala, N. (2022). Integrating UUVs for naval applications. Maritime Technology and Research, 4(3), 254470. https://doi.org/10.33175/mtr.2022.254470

Benner, T. (2025, January 8). The Underwater Battlefield: Protecting Submarine Critical Infrastructure. Internationale Politik Quarterly. Retrieved from: https://ip-quarterly.com/en/underwater-battlefield-protecting-submarine-critical-infrastructure

BMT. (2024, July 11). BMT and Ocius to Revolutionise Maritime Data Collection with Autonomous "Satellites of the Sea". Retrieved from: https://www.bmt.org/news/2024/bmt-and-ocius-to-revolutionise-maritime-data-collection-with-autonomous-satellites-of-the-sea/

CCDCOE. (2019, November). Strategic importance of, and dependence on, undersea cables. NATO Cooperative Cyber Defence Centre of Excellence. Retrieved from: https://ccdcoe.org/uploads/2019/11/Undersea-cables-Final-NOV-2019.pdf

Jensen, V. (2022). *Deep, Dark and Dangerous: The Story of British Columbia's World-class Undersea Tech Industry*. Harbour Publishing Company

Kajal, K. (2024, August 5). Nezha-SeaDart: China tests air-sea hybrid drone with flight, dive capabilities. Retrieved from *Interesting Engineering*: https://interestingengineering.com/innovation/china-drone-air-underwater-hybrid

Martin, B., Tarraf, D. C., Whitmore, T. C., DeWeese, J., Kenney, C., Schmid, J., & DeLuca, P. (2019). *Advancing Autonomous Systems: An Analysis of Current and Future Technology for Unmanned Maritime Vehicles*. RAND Corporation. California: RAND Corporation. doi:10.7249/RR2664

Mathe, S.E., Kondaveeti, H.K., Vappangi, S., Vanambathina, S.D., Kumaravelu, N.K. (2024). A comprehensive review on applications of Raspberry Pi. *Computer Science Review*, 52, ISSN 1574-0137. Retrieved from: https://www.sciencedirect.com/science/article/pii/S1574013724000200

Mishra, P.R. (2025, June 1). US proves battle readiness, deploys robotic, autonomous submarine from torpedo tubes. Retrieved from Interesting Engineering: https://interestingengineering.com/military/us-deploys-robotic-submarine

Sciutto, J. (2024, September 6). US sees increasing risk of Russian 'sabotage' of key undersea cables by secretive military unit. CNN Politics. Retrieved from: https://edition.cnn.com/2024/09/06/politics/us-sees-increasing-risk-of-russian-sabotage-undersea-cables/index.html

Willett, L. (2024, July 9). First US Navy Submarine Will Deploy with New UUV Capability this Year. Retrieved from Naval News: https://www.navalnews.com/naval-news/2024/07/first-us-navy-submarine-will-deploy-with-new-uuv-capability-this-year/

Index